DATE DUE

WITHDRAWN
UTSA LIBRARIES

GAME THEORY AS A THEORY OF CONFLICT RESOLUTION

THEORY AND DECISION LIBRARY

AN INTERNATIONAL SERIES
IN THE PHILOSOPHY AND METHODOLOGY OF THE
SOCIAL AND BEHAVIORAL SCIENCES

Editors:

GERALD EBERLEIN, *Universität des Saarlandes*
WERNER LEINFELLNER, *University of Nebraska*

Editorial Advisory Board:

K. BORCH, *Norwegian School of Economics and Business Administration*
M. BUNGE, *McGill University*
J. S. COLEMAN, *Johns Hopkins University*
W. KROEBER-RIEL, *University of Saarland*
A. C. MICHALOS, *University of Guelph*
A. RAPOPORT, *University of Toronto*
A. SEN, *University of London*
W. STEGMÜLLER, *University of Munich*
K. SZANIAWSKI, *University of Warsaw*
L. TONDL, *Prague*

VOLUME 2

GAME THEORY AS A THEORY OF CONFLICT RESOLUTION

Edited by

ANATOL RAPOPORT

University of Toronto

D. REIDEL PUBLISHING COMPANY

DORDRECHT-HOLLAND / BOSTON-U.S.A.

Library of Congress Catalog Card Number 73-91434

Cloth edition: ISBN 90 277 0424 4
Paperback edition: ISBN 90 277 0489 9

Published by D. Reidel Publishing Company,
P.O. Box 17, Dordrecht, Holland

Sold and distributed in the U.S.A., Canada and Mexico
by D. Reidel Publishing Company, Inc.
306 Dartmouth Street, Boston,
Mass. 02116, U.S.A.

All Rights Reserved
Copyright © 1974 by D. Reidel Publishing Company, Dordrecht, Holland
No part of this book may be reproduced in any form, by print, photoprint, microfilm,
or any other means, without written permission from the publisher

Printed in The Netherlands by D. Reidel, Dordrecht

LIBRARY
University of Texas
At San Antonio

CONTENTS

ANATOL RAPOPORT / Introduction ... 1

PART I. TWO-PERSON GAMES

ANATOL RAPOPORT / Prisoner's Dilemma – Recollections and Observations ... 17

T. BURNS and L. D. MEEKER / Structural Properties and Resolutions of the Prisoners' Dilemma Game ... 35

D. MARC KILGOUR / On 2×2 Games and Braithwaite's Arbitration Scheme ... 63

CLEMENT S. THOMAS / Design and Conduct of Metagame Theoretic Experiments ... 75

ANATOL RAPOPORT and J. PERNER / Testing Nash's Solution of the Cooperative Game ... 103

PART II. N-PERSON GAMES

JAMES P. KAHAN and AMNON RAPOPORT / Test of the Bargaining Set and Kernel Models in Three-person Games ... 119

ABRAHAM D. HOROWITZ and AMNON RAPOPORT / Test of the Kernel and Two Bargaining Set Models in Four- and Five-person Games ... 161

D. MARC KILGOUR / A Shapley Value for Cooperative Games with Quarrelling ... 193

JAMES D. LAING and RICHARD J. MORRISON / Coalitions and Payoffs in Three-person Supergames under Multiple-trial Agreements ... 207

M. FREIMER and P. L. YU / The Application of Compromise Solutions to Reporting Games ... 235

NIGEL HOWARD / 'General' Metagames: An Extension of the Metagame Concept ... 261

ANATOL RAPOPORT

INTRODUCTION

Game theory could be formally defined as a theory of rational decision in conflict situations. Models of such situations, as they are conceived in game theory, involve (1) a set of decision makers, called *players*; (2) a set of *strategies* available to each player; (3) a set of *outcomes*, each of which is a result of particular choices of strategies made by the players on a given play of the game; and (4) a set of *payoffs* accorded to each player in each of the possible outcomes.

It is assumed that each player is 'individually rational', in the sense that his preference ordering of the outcomes is determined by the order of magnitudes of his (and only his) associated payoffs. Further, a player is rational in the sense that he assumes that every other player is rational in the above sense. The rational player utilizes knowledge of the other players' payoffs in guiding his choice of strategy, because it gives him information about how the other players' choices are guided.

Since, in general, the orders of magnitude of the payoffs that accrue to the several players in the several outcomes do not coincide, a game of strategy is a model of a situation involving conflicts of interests.

A player may choose his strategies probabilistically, that is, by means of a random device that assigns fixed probabilities to his own strategy choices. A combination of such probabilistically chosen strategies, called *mixed* strategies, determines a probability distribution of the outcomes, hence also of the expected payoffs accruing to each player. In that case, a player's decisions are guided by an attempt to maximize his expected payoff. The justification of this assumption derives from the definition of utility, in terms of which the payoffs are presumably given. Strategies that are not mixed are called *pure*.

Games of strategy can be classified with reference to the number of players, the important distinction being between 2-person games and N-person games ($N > 2$), and according to whether the choices of strategies must be made independently (non-cooperative games) or may be coordinated (cooperative games). Another important distinction is between constant-sum games, where the sum of the payoffs to the players is the same in each outcome, and non-constant-sum games. In 2-person constant-sum games, the interests of the players are necessarily diametrically opposed, since the greater the payoff to one of the players, the smaller it must be to the other. In non-constant-sum games, the interests of the players may partially coincide. In N-person games, even if they are constant-sum, the interests of some of the players may also partially coincide. It is such situations – where the interests of participants partially conflict and partially coincide – that occur most frequently in human conflicts.

The origins of game theory are rooted in the analysis of the 2-person constant-sum game, a model of an irreconcilable conflict. In fact, 2-person games of strategy, so called parlor games, e.g., Chess, Go, etc., are models of conflicts of this sort. Skill in such games depends on the ability of finding 'good strategies', those that lead to one's own advantage and consequently to the disadvantage of the co-player.

The fundamental theorem of game theory, proved by von Neumann (1928), asserts that every 2-person constant-sum game with a finite number of strategies has a solution in the following sense: there exists, available to each player at least one *optimal* strategy, which may be pure or mixed. A player choosing such an optimal strategy can guarantee himself a certain minimal payoff (or minimal expected payoff). By definition of a constant-sum game, this means that each player can keep the other's payoff down to the latter's guaranteed minimum. The resulting outcome of the game is an *equilibrium*. That is, if one player keeps to his optimal strategy, the other cannot improve his payoff by shifting away from *his* optimal strategy; and he may impair it.

Further, it is shown in the theory of the 2-person constant-sum game that if there are several such equilibria, determined by paired choices of optimal strategies, then they are all *equivalent* and *interchangeable*. They are equivalent in the sense that the payoffs to each player are the same in all of them. They are interchangeable in the sense that if each player chooses an arbitrary strategy among whose possible outcomes is an equilibrium, then the result of the pair of choises will always be an equilibrium.

Equivalence and interchangeability of equilibria make it possible to view the 'solution' of the 2-person constant-sum game as a *normative* one.

It is possible to *prescribe* an optimal strategy to each player, namely, by advising him to choose any of the strategies that contains an equilibrium among its outcomes. If both players follow this prescription, each will do as well as he possibly can in that game (against a rational player). Moreover, since the payoffs in all equilibria are equal, it does not matter which of the optimal strategies he chooses. In this context, therefore, game theory well deserves the designation as a theory of rational (i.e., normatively prescribable) decisions in conflict situations.

Attempts to extend this approach to non-constant-sum games and to games with more than two players are beset with difficulties. To be sure, the existence theorem cited above has been extended. It is now known that every non-cooperative game with a finite number of players and strategies has at least one equilibrium (Nash, 1950). However, if such a game has several equilibria, they are not necessarily equivalent or interchangeable. Therefore it is not possible to extend the prescription, 'Choose a strategy containing an equilibrium', to games that are not 2-person constant-sum games. As an example, consider the game represented by Matrix 1, where player 1 chooses horizontal rows, player 2 vertical columns, and the entries are respectively to player 1 and player 2.

	A_2		B_2	
A_1	3,	3	2,	4
B_1	4,	2	1,	1

Matrix 1

Both of the outcomes A_1B_2 and B_1A_2 are equilibria, since if either player 'shifts away' from either of those outcomes, he thereby impairs his own payoff. Now if a player is advised to 'choose a strategy that contains an equilibrium', player 1 may well choose B_1, because he prefers the equilibrium at B_1A_2 to that at A_1B_2; and player 2 may choose B_2 for a similar reason. The resulting outcome will be B_1B_2, which is not an equilibrium. In fact, any of the four outcomes may result if each player chooses a strategy containing an equilibrium.

Even if the equilibria of a non-constant-sum game are all equivalent and interchangeable, so that the abovementioned difficulty cannot occur, the identification of an equilibrium outcome with the result of 'rational decisions' is open to objections. Consider the game represented by Matrix 2.

	A_2	B_2
A_1	3, 3	1, 4
B_1	4, 1	2, 2

Matrix 2

This game contains only one equilibrium, namely at $B_1 B_2$. Therefore, if each player is advised to 'choose a strategy containing an equilibrium', the equilibrium at $B_1 B_2$ will obtain. But *both* players would have got a larger payoff in $A_1 A_2$ than in $B_1 B_2$. Can decisions be defended as rational if they result in an outcome that is worse for both players than a different one? The paradoxical 'solution' suggests that the concept of 'rationality' should be re-examined, perhaps split into two concepts, individual rationality and collective rationality. The outcome of a non-constant-sum game may be dictated by the individual rationality of the respective players without satisfying a criterion of collective rationality.

Collectively rational outcomes are at the center of attention in the theory of cooperative games, in which players can make binding agreements and choose their strategies *jointly*. For instance, in the game represented by Matrix 2, the two players can agree to choose A_1 and A_2 respectively and so assure the collectively rational outcome. Clearly in 2-person *constant-sum* games, joint decisions serve no purpose, because in such games there are no outcomes that are preferred by both players to others. In games with more than two players, however, a community of interest may exist among some of the players, even if the game is contant-sum, because they may be able to get jointly more if they coordinate their strategies, so as to keep the joint payoff of the remaining players to a minimum.

A set of players that coordinate their strategies in order to promote their joint interest is called a *coalition*. *N*-person game theory involves for the most part, questions related to the formation of coalitions and the apportionment of their jointly gained payoffs among their members. In this way, both the collective interests and the individual interests of the players enter the theory. To the extent that the players act so as to get the largest joint payoff, they are guided by their collective interest. When they face the problem of apportioning their joint gain among themselves, each player bargains for his share, and individual interests come into play. The theory of the N-person cooperative game examines the bargaining positions of the players in terms of their potential contributions to the several coalitions that may form. In this phase of the game, a player's resources are in the form of threats and promises that he can make to join or to withdraw from some coalition.

If a cooperative N-person game is examined from this point of view, the two cardinal questions are (1) How will a set of players of a given game partition themselves into coalitions? and (2) How will the members of each coalition apportion their joint payoff among them? These questions can be viewed in the context of a descriptive (or predictive) theory, subject to experimental corroboration, or in the context of a prescriptive (normative) theory. In the latter context, the first question above is easily answered: all N players should form the 'grand' coalition; for this inclusive coalition can always get *jointly* at least as much as several disjoint coalitions. The other problem, however, that of apportioning the jointly gained payoff among the members of the grand coalition remains.

Several solutions of this problem have been proposed. The so called von Neumann-Morgenstern solution, advanced in the original treatise on game theory (von Neumann and Morgenstern, 1947), singles out certain *sets* of apportionments, called *imputations*, each set being *a* solution of the given game. The rationale behind this concept of solution is that each of those sets possesses a 'stability' of sorts. Nevertheless, the von Neumann-Morgenstern solutions leave much to be desired, inasmuch as the sets comprising each solution generally involve infinite sets of imputations, and moreover a game may have an infinity of solutions. Besides, Lucas (1968) has shown that some N-person games have no von Neumann-Morgenstern solutions.

Shapley (1953) has advanced a method of singling out a single imputation of an N-person cooperative game, now called the *Shapley value* of the game. This 'solution' can be supported by a rationale that leans more on considerations of equity than of stability.

If game theory is viewed as a tool of research in behavioral science, that is, as a descriptive or predictive rather than a normative theory, then the assumption that the players of an N-person game will form a grand coalition can no longer be made; for often they do not. In some treatments of N-person game theory, the question of which coalitions will actually form is side-stepped. It is assumed simply that the players will *somehow* partition themselves into coalitions. *Given* a particular coalition structure, whereby each coalition obtains exactly the minimum joint payoff guaranteed to it by the structure of the game, the problem now is how to predict (or prescribe) the apportionment of payoffs within each coalition, taking into account the pressures that players can exert on each other. (These pressures derive from possible re-alignments.) This approach leads to the concept of the *bargaining set* (Aumann and Maschler, 1964), the *kernel* (Davis and Maschler, 1965), the *competitive bargaining set* (Horowitz, 1972), and others, each set being a set of payoff apportionments coupled with specific coalition structures, such that the two together possess certain stability properties.

In all these developments of the theory of the N-person cooperative game, the concept of 'optimal strategy', which had been central in the theory of the 2-person constant-sum game, has lost its salience. The starting point of the theory is now the assumption that any individual player or any set of players in a coalition *can* find a strategy that will guarantee them some minimum payoff. The focus of the theory is thereby shifted to an entirely different class of problems, namely, those that arise in the interplay of the interests of individual players, those of proper subsets of the set of N players, and the collective interests of all N players. The proposed 'solutions' are stated not in the form of 'how to play the game so as to get most' but in terms of balancing all of these criss-crossed pressures and interests. If such solutions are offered as 'rational' *final outcomes* of the game, game theory becomes primarily a theory of conflict resolution rather than a theory of optimal decisions in the pursuit of self interest. It is, perhaps, inevitable that game theory should have taken this direction, because, as we have seen, unambivalent prescriptions of optimal strategy choices made to individual players are no longer possible, once the format of the 2-person constant-sum game has been transcended.

In view of this development, game theory can be said to have disappointed some hopes but also to have stimulated others. The hopes that have been disappointed (or should have been disappointed if the findings of the theory are properly understood) were those of using game-theoretic analysis to generate techniques of policy construction in typically occurring conflict situations, such as war, war planning, power politics, and business competition. Since these fields of activity are prominent and prestigious in 'advanced' societies, the demand for strategic expertise is large, and the lively interest in game theory, manifested now and then among professionals engaged in those activities, is understandable. And so is the disappointment engendered by the lack of success in linking formal game theory to competitive expertise. The usual explanation of this failure, to the effect that the real world is too complex to be simulated by formal game-theoretic models, misses the main point: even if real conflicts were no more complex than theoretically tractable formal games, game theory would be powerless to prescribe 'optimal strategies' in any but total bi-polar conflicts for reasons stated above.

The hopes that were stimulated point in another direction, that of re-formulating the problems generated by conflict situations. In differentiating 2-person constant-sum games from all others, game theory draws a sharp line between conflicts that are clearly irreconcilable and those that admit outcomes preferred by all participants to other outcomes resulting from independent pursuits of individual interests. In singling out these so called *Pareto-optimal* outcomes, the theory of the cooperative N-peron game poses the problem of conflict resolution, inasmuch as with regard to the choice among the several Pareto-optimal outcomes and the apportionment of joint payoffs, the interests of the players do conflict. However (and this is where game theory is most relevant to a theory of human conflict), even in jockeying for advantageous positions on the Pareto-optimal set of outcomes, 'rational' players must continue to keep collective interests in mind if outcomes bad for all are to be avoided.

It goes without saying that if game theory is to make an input to behavioral science, experimental and theoretical developments must go on apace. However the two need not be directly linked in every instance. Purely formal results of the theory can and often do suggest new questions to be answered by experimental procedures. Unexpected experimental results or even real life observations may instigate new lines of theoretical development.

Part I of this volume deals with the 2-person non-constant-sum game. Games of this sort, most notably one called Prisoner's Dilemma (represented in Matrix 2), have become instruments of laboratory studies of 'conflict', induced by the divergent interests of two players and mitigated by their common interest. The first article includes a brief overview of the experimental work with Prisoner's Dilemma and of its impact on the theoretical and experimental approach to the study of conflict. Three theoretical papers follow. One contains reports of some experiments as illustrations of a method. The last paper in Part I is on a *cooperative* two-person game which is *not* Prisoner's Dilemma. The latter paper was included, because experiments with cooperative two-person games are still quite rare. This may seem strange at first sight, because definitive game-theoretic solutions for the cooperative game have been proposed, whereas concepts of 'solutions' of non-cooperative games are beset with ambiguities. The heavy emphasis on the non-cooperative game in experimental literature is explained by the fact that it was primarily behavioral scientists (in some cases students of management science) that have created the field of experimental games, not game theoreticians. The latter, being primarily concerned with the mathematical ramifications of game theory are not, as one would expect, inclined to view game theory from the behavioral point of view and to put its findings to an experimental test.

Of the three theoretical papers, two (Thomas' and Kilgour's) are mathematical-formal. That is, they present arguments and conclusions exclusively in accordance with the rules of strict deduction. Consequently, the mathematical symbolism in these papers is a necessary, at times indispensable tool of reasoning. Obviously, some one unable to read the symbolism cannot follow the reasoning that leads to the conclusions. Whether or not the reasoning and the conclusions could be explained without recourse to at times formidable notation with its mixture of Roman and Greek alphabets and several fonts bristling with subscripts and superscripts, is a question that can be answered only in specific cases. *In general*, mathematical reasoning makes mathematical notation necessary. The conclusion is unescapable that sooner or later those interested in mathematico-logical analysis of the logical structure of conflict (which is what game theory is about) will have to become familiar with this bizarre language. In the other theoretical paper (Burns and Meeker's) the authors also make extensive use of mathematical notation. It is not a mathematical paper, however, because the notation is not used to make mathematical deductions. It only serves to make verbal definitions precise. The arguments themselves remain on the verbal level, the traditional mode of discourse in the social sciences.

I confess to feelings of strong ambivalence with regard to the use of elaborate mathematical symbolism in a discussion where no use is made of mathematical deductions. I have read innumerable papers in this genre and could not suppress the suspicion that the symbolism serves as window dressing to camouflage paucity or triviality of content. On the other hand, I feel strongly that social scientists *should* learn the language of mathematics. If the only way they can learn it is by translating plain English into math ('new' math, at that – the set-theoretic dialect), so be it. After all, precise definitions (which are guaranteed by mathematically grammatical notation) are a first step toward mathematical reasoning. To be sure, some social scientists have never got beyond that first step – setting up the 'system', definitions, axioms, etc. – 'getting ready to get ready'. But some have taken off and thereby have broken new ground in social science; so there is hope.

In spite of my misgivings concerning the use of mathematical symbolism in Burns and Meeker's paper, which (notation) both the mathematical and the non-mathematical reader may safely skip, I have selected that paper for Part I of this volume, because it is a representative sample of the sort of thinking stimulated in social scientists by the much publicized Prisoner's Dilemma (see the remarks in my own article on the subject). Note that from the point of view of the *game theoretician* Burns and Meeker are breaking through an open door. What they are in effect saying is that payoffs in any formally presented game, in particular Prisoner's Dilemma, do not actually represent the players' utilities. Players are often concerned not only with their own payoffs but also with what the co-player gets, sometimes empathizing with him, sometimes, on the contrary, deriving satisfaction from his losses, regardless of what they themselves get. Real players play differently depending on whom they are playing with or against. Learning involves not only a recognition of the strategic structure of a game but also modification of one's preferences (that is a different translation of payoffs into utilities), etc. All these observations, while perfectly true, are of absolutely no concern to the game-theoretician. The game theoretician is concerned with the mathematical model and with conclusions he can draw from assumptions (in particular about utilities represented by payoffs) *that stay put*, not with assumptions that capture the entire spectrum of human conflict behavior. Nevertheless the notion that 'game theory' is a theory of behavior (albeit idealized) persists. Thus, controversies about the extent of 'relevance' of game-theoretic conclusions to real conflict behavior are based on a misconception. What *should* be kept in mind is that in *any* situation involving conflicts and decisions the overtly assigned payoffs are not necessarily the payoffs that guide the decisions. For this reason, it is all but useless to assess the 'rationality' of any observed course of action except with reference to pre-conceived notions of utilities, fixated as overt payoffs, which may or may not reflect the essentials of the situation.

Part II dealing with N-person games contains two experimental and three theoretical papers. In the first two papers, specific proposed solutions of N-person games are tested experimentally. In the 3-person games examined, the bargaining set and the kernel coincide; so that the two solutions are tested together. These are 'point' solutions: that is, they prescribe unique apportionments of payoffs for each coalition structure. They do not, however, prescribe specific coalition structures, so that the formation of specific coalitions must be ascribed to factors not taken into account in the models in question. In particular, experimental conditions, specific magnitudes of payoffs, and learning may be among the factors influencing both the coalition structure and the apportionments. To the extent that these factors may involve 'psychological' components, they are properly the domain of behavioral theory rather than of the formal theory of games, in which the players are not endowed with any 'psychology' except that they are assumed to be 'rational' in the sense defined above. Since real players all have 'psychologies', any attempt to link game-theoretic formulations with experimental procedures must involve a fusion of game theory and some behavioral theory. The first two papers are examples of such integration.

Kahan and Rapoport observe that in the 3-person game, the most frequent coalition is that between the two stronger players against the weakest and remark that this result is at variance with the social scientific theories predicting a predominance of coalitions of the weak against the strong. It should be pointed out that the definitions of 'strong' and 'weak' players are not always consistent. In some experiments reported by Gamson (1961), it seems that indeed the 'weak' tend to combine against the 'strong'. For example, if a decision is to be made by majority vote and if three parties have 40, 35, and 25 votes respectively, the two 'weaker' parties are more likely to combine against the 'strong' one than the two stronger against the weaker one. In this situation, winner takes all. In the game considered by Kahan and Rapoport, the situation is different. Each pair coalition is guaranteed a minimum payoff of a different magnitude. The 'strongest' player is one who in coalition with each of the other two assures the coalitions the two larger amounts; the 'weakest' player is one who in coalition with each of the others assures the coalitions the two smaller amounts. Thus, in this situation there is an inducement for the two stronger players to join in a coalition, which is absent in the 'winner takes all' game. In fact, in the latter case, there may be a stronger inducement for the two weaker players to join in order to avoid the strongest player's claim to a larger share of the joint payoff. It is noteworthy that an analysis of the actual bargaining strength of the three players in a winner-take-all game, the 'numbers of votes' (or 'resources', as these are sometimes called in experiments of this sort) may have no bearing on the structure of the game, since the size of the majority is here irrelevant. They may, of course, be psychologically relevant; so that the principal value of formal analysis is to separate real strategic considerations from 'psychological' ones.

In the second paper, Horowitz and Rapoport extend their method to 4- and 5-person games. Here the bargaining set comprises not unique apportionments but only ranges in which they must fall. It turns out that the kernel of the game is at one extreme of such a range, while the competitive bargaining set is at the other. Thus the kernel and the competitive bargaining set can be pitted against each other in an experiment. It is interesting to note that the experimental evidence definitely favors the latter over the former. Another noteworthy finding is that the trend of outcomes with accumulating experience is away from more 'egalitarian' apportionments to outcomes in which the relative bargaining positions of the players become more salient.

Laing and Morrison develop two variants of a behavioral theory aimed at predicting (probabilistically) which coalitions will form in 3-person games played sequentially. The payoffs in these games serve as incentives only indirectly, via their influence on the rank achieved by a player after an unspecified number of plays. Thus, an extra competitive component is introduced: the players 'race' for rank position at the end of the sequence. In Laing and Morrison's theory, this component is the principal determinant of the relative attractiveness of the players for each other as coalition partners, hence of the relative frequencies of pair coalitions. As the authors point out, the situation depicted is "sufficiently complex that neither players nor analysts are able to represent [it] as a whole in a form amenable to game theoretic solution." The authors, therefore, take recourse in a relatively simple behavioral theory, as do many experimenters interested in the way people actually tend to behave in conflict situations.

The next three papers are purely theoretical. Kilgour generalizes the concept of Shapley value of an N-person game to the case where the grand coalition cannot form because of 'quarrels' involving certain players, which make it impossible for them to belong to the same coalition. The basic idea here is that of introducing 'friction' into the social process, thus modifying the 'classical' theory, which might be conceived as one that examines 'depsychologized' conflict, in vacuo, as it were. The interesting finding here is that whereas under certain formulations of the rules of coalition formation, quarrels are always to the disadvantage of the quarrelers, as, perhaps, would be expected, under other formulations, quarrels may be actually to the advantage of the quarrelers. Implications suggested by this somewhat unexpected finding can be readily imagined.

The paper by Freimer and Yu begins with the problem of arriving at compromise group decisions. The solution is offered in terms of minimizing so called *group regret*, defined as a generalized distance from the point representing each group member's utility, associated with the decision actually made, to the so called *utopia point*, where every one can have his way. Of course in the real world the utopia point is usually unattainable, hence the necessity of compromise. The interesting feature is the parametrization of the distance function. As is well known, in Euclidean n-dimensional space, the distance between the point $(x_1, x_2, ..., x_n)$ and the point $(x'_1, x'_2, ..., x'_n)$ is defined as $[\sum_{i=1}^{n}(x_i - x'_i)^2]^{1/2}$. In the generalized definition, the exponent 2 is replaced by a parameter $p (1 \leq p < \infty)$. This parameter has an interesting meaning. The closer it is to 1, the more weight does the group regret have over the individual regret; the larger it is, the more weight does the largest of the individual regrets have over the group regret. Thus, the parameter p is a sort of measure of the extent to which concern for how strongly the members of a group feel about the issues in question is manifested in the group decision norm. In principle, this parameter could be estimated in group decision experiments and so could provide bases of comparison between groups or populations from which they are recruited.

The theory of group decisions has only a tangential connection to game theory, since in the normative aspect of the former, bargaining leverage does not play the strong part it plays in the N-person game. For instance, Freimer and Yu's compromise solution is based on what amounts to an equity principle rather than on the assessment of the power position of the players. In the extreme case, when p is very large, the individual who feels 'most strongly' about the issues has his way. Nevertheless, genuine game theory enters the formulation by the back door, as it were. For if 'feeling strong about the issue' confers an advantage, individuals may give false reports about their utilities for the various outcomes. A deterrent against falsifying these reports is the danger of 'losing credibility' next time around, and so the problem of finding optimal reporting strategies in what now becomes a non-cooperative game again comes to the forefront, and the theory has gone through a complete cycle.

The last paper by Howard extends metagame theory, previously developed by the same author. Knowledge of that theory is presupposed in the paper. The interested reader is referred to the book on the subject (Howard, 1971). Metagame theory comes, perhaps, closest to the central theme of this book; for, as the author points out, it represents essentially a fusion of the theories of non-cooperative and cooperative games. The fusion is achieved by passing from the space of strategies that define a game in normal form to a space of metastrategies, defined as the strategies that would be available to a player if he knew the strategy choices of the the other players. It turns out that outcomes of metagames singled out as stable in the game-theoretic sense include some which, although not stable in a non-cooperative game, can be solutions of a cooperative game. In this way, metagame theory can be viewed as a theory of conflict resolution by way of 'stretching' the conceptual repertoire of the participants.

Dept. of Psychology, Univ. of Toronto

BIBLIOGRAPHY

Aumann, R. J. and Maschler, M., 'The Bargaining Set for Cooperative Games' in *Advances in Game Theory* (ed. by M. Dresher, L. S. Shapley, and A. W. Tucker),

Annals of Mathematics Studies **52** Princeton University Press, Princeton, N.J., 1964.

Davis, M. and Maschler, M., 'The Kernel of a Cooperative Game', *Naval Research Logistics, Quarterly* **12** (1965) 223–59.

Gamson, W. A., 'A Theory of Coalition Formation', *American Sociological Review* **26** (1961) 373–82.

Horowitz, A. D., 'The Competitive Bargaining Set for Cooperative n-Person Games', *Journal of Mathematical Psychology* **10** (1973) 265–289.

Howard, N., *Paradoxes of Rationality*, MIT Press, Cambridge, Mass., 1971.

Lucas, W. F., 'The Proof that a Game May not Have a Solution', Memorandum RM–5543–PR, January 1968. The Rand Corporation, Santa Monica.

Nash, J. F., 'Equilibrium Points in n-Person Games', *Proceedings of the National Academy of Sciences, U.S.A.* **36** (1950) 48–49.

Shapley, L. S., 'A Value for n-Person Games', in *Contributions to the Theory of Games*, Vol. II, (ed. by H. W. Kuhn and A. W. Tucker), *Annals of Mathematics Studies* **28** Princeton University Press, Princeton, N.J., 1953.

Von Neumann, J., 'Zur Theorie der Gesellschaftsspiele', *Mathematische Annalen* **100** (1928) 295–320.

PART I

TWO-PERSON GAMES

ANATOL RAPOPORT

PRISONER'S DILEMMA – RECOLLECTIONS AND OBSERVATIONS

Two prisoners accused of the same crime are kept in separate cells. Only a confession by one or both can lead to conviction. If neither confesses, they can be convicted of a lesser offense, incurring a penalty of one month in prison. If both plead guilty of the major crime, both receive a reduced sentence, five years. If one confesses and the other does not, the first goes free (for having turned State's evidence), while the other receives the full sentence, ten years in prison. Under the circumstances is it rational to admit guilt or to deny it?

If my partner confesses (so each prisoner might reason), I stand to gain by confessing, for in that case, I get five years instead of ten years, if I don't confess. If, on the other hand, my partner does not confess, it is still to my advantage to confess, for a confession sets me free, while otherwise I must serve a month. Therefore I am better off confessing *regardless* of whether my partner does or not.

The 'partner', being in the same situation, reasons the same way. Consequently, both confess and are sentenced to five years. Had they not confessed, they would have been sentenced to only one month. In what sense, therefore, can one assert that 'to confess' was the prudent (or 'rational') course of action?

The anecdote is attributed to A. W. Tucker, and the game depicting the situation has been christened appropriately 'Prisoner's Dilemma'. It is an example of a two-person non-constant-sum game, one in which some outcomes are preferred by *both* players to other outcomes. The dilemma arises from the circumstance that in the absence of communication and hence of making binding agreements, there is no way of rationalizing the choice of action, which if taken by both players benefits both.

Many situations analogous to Prisoner's Dilemma present themselves in real life. Two rival powers may be equally 'safe' against each other if both disarm or if both remain armed. Moreover, the disarmed state is cheaper, so that both would gain by disarming. Yet it is to each power's individual advantage to remain armed. For being armed against a dis-

Anatol Rapoport (ed.), Game Theory as a Theory of Conflict Resolution, 17–34. All Rights Reserved. Copyright © 1974 by D. Reidel Publishing Company, Dordrecht-Holland

armed adversary is 'obviously' an advantage, while being disarmed and facing an armed adversary is 'unthinkable'. (The quotation marks indicate that the perceptions are those of contemporary conventional wisdom.)

Or, consider two competing firms selling the same product and faced with a choice of charging a high or a low price. To undersell the competitor is an advantage; to be undersold by him is disastrous. So both sell at a low price; whereas both would make more profit if both sold at a high price. Price fixing is, of course, a common practice, but it depends on explicit or tacit agreements. The dilemma arises when such agreements are impossible or cannot be enforced. In the latter case, the choice is between keeping or breaking the agreement. Breaking it is again to each competitor's advantage.

The dilemma in Prisoner's Dilemma arises from the circumstance that the question 'What is the rational choice?' is ambivalent unless 'rationality' is strictly defined. It turns out that in the context of non-constant-sum games like Prisoner's Dilemma actually two concepts of 'rationality' compete for attention, namely *individual* rationality, which prescribes to each player the course of action most advantageous to him under the circumstances, and *collective* rationality, which prescribes a course of action to both players simultaneously. It turns out that if both act in accordance with collective rationality, then *each* player is better off than he would have been had each acted in accordance with individual rationality. The situation has been, of course, known and intuitively appreciated throughout man's history as a social animal. Prescriptions of conduct in accordance with collective rationality are embodied in every ethical system. Such a prescription is spelled out in Kant's categorical imperative. It is incorporated in every disciplined social act, for instance in an orderly evacuation of a burning theater, where acting in accordance with 'individual rationality' (trying to get out as quickly as possible) can result in disaster for all, that is, for each 'individually rational' actor.

Nevertheless the formal presentation of the Prisoner's dilemma game still evokes surprise in people confronted with the analysis for the first time. Perhaps this is because acting in collective interest is still associated in most people's minds with moral rather than logical concepts. Doing what is for the common good evokes the idea of 'sacrificing' one's individual interest. On the other hand, calculation of what is best for oneself is felt to be unambivalently 'rational' (although perhaps on occasions not praiseworthy). There is a reluctance to question the very meaning of 'rationality', as one must when one becomes aware of the implications of purely logical analysis when applied to games like Prisoner's dilemma.

Accordingly, a controversy is still carried on here and there about whether the one or the other choice in that game is 'really' rational. What is 'really' rational is, of course, a matter of definition and so not something that deserves a serious controversy. Much more interesting is the question of what people actually do when confronted with a Prisoner's Dilemma situation. In fact, if we could get an independent assessment of what constitutes a 'rational actor', the question of which choice is rational in Prisoner's Dilemma could be put to an experimental test. Then what a 'rational actor' does will provide the answer to the question.

Unfortunately, it is manifestly impossible to agree on criteria that designate a 'rational actor' to apply to real flesh and blood actors. However the question 'What will people (any people) do in a Prisoner's Dilemma situation?' remains an interesting one, as is attested by several hundreds of experiments that have been performed with this game in the past twenty years or so. This research is part of the 'history' of Prisoner's Dilemma. Another part is the penetration of the 'model' into discussions of international relations and of other situations where the psychology of conflict resolution seems to be potentially appropriate. For Prisoner's Dilemma is not only a model of a peculiar sort of conflict, involving 'partner-opponents', whose interests partly clash and partly coincide but also an inner conflict within each player: 'Shall I cooperate (act the way I want the other to act) or compete (look after my own interest)?' It is therefore not surprising that the game acquired a strong appeal for experimenting psychologists.

To my knowledge, the earliest experiments with Prisoner's Dilemma were performed by Flood in 1952. They were reported in a Rand Corporation research memorandum [1] and do not seem to have attracted much attention at the time. The 'paradox' was discussed by several of the Fellows at the Center for Advanced Study in the Behavioral Sciences in Palo Alto during the first year of its operation (1954–1955). I believe it was introduced by Luce, and the first detailed analysis appeared subsequently in his and Raiffa's *Games and Decisions* [2]. For reasons I will mention below, those discussions made a very deep impression on me, and I carried the 'infection' with me to the Mental Health Research Institute at the University of Michigan, where I worked from 1955 to 1970. However, my own systematic experiments with Prisoner's Dilemma did not begin until 1962.

Possibly a decisive impetus to experimental work was given by a paper by Schelling [3], published in 1958. At any rate, it seems that the first experiment since Flood's was performed by Deutsch in 1958 [4]. Thereafter the number of experimental papers on Prisoner's Dilemma increased very rapidly.

The growth of interest in that game can be estimated from the time course of papers on experimental games and related subjects of which there are several comprehensive bibliographies [5]. The number of papers published per year in twenty years cited in the Guyers and Perkel bibliography is plotted in Figure 1. Of these, a large majority are reports of experiments; and of the latter at least one-half involve the standard Prisoner's Dilemma game in matrix form. I estimate that at least 200 experiments with Prisoner's Dilemma have been reported, the bulk of

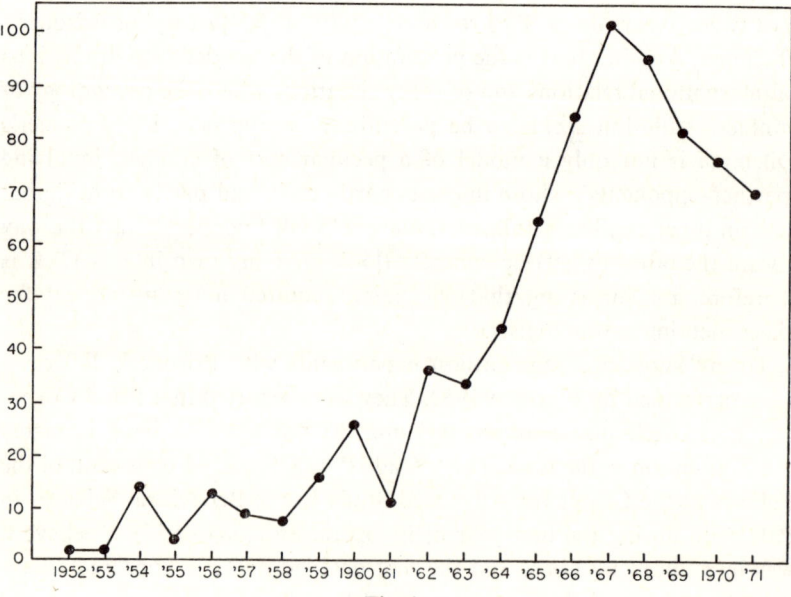

Fig. 1.

them since about 1965. In trying to account for this minor explosion, I shall resort to a bit of personal history.

My first paper involving a mathematical model of interactive behavior was published in 1947 [6]. It was an elaboration of a scheme used by my mentor Nicolas Rashevsky, a pioneer in mathematical biology, mathematical psychology, and mathematical sociology. The model is that of two individuals, each producing some product and sharing it in a fixed ratio, $p:q$. The activities of the individuals are governed by a 'satisfaction function' (utility) which is a sum of two terms. The first is a positive logarithmic function of the amount of product received; the second is negative, proportional to the effort expended. Thus, if an individual appropriated all of the product, he would produce an amount that maximized his utility. When the two produce jointly and share, each might try to maximize his own utility. In general, there is an equilibrium at which both utilities are *locally* maximized. Depending on the parameters of the model (the coefficients of proportionality and the ratio $p:q$), this equilibrium may be stable or unstable. If it is unstable, a slight decrease of effort (rate of production) on the part of one individual stimulates an *increase* of effort on the part of the other, and this tendency continues until one of the individuals produces nothing, that is, becomes a 'parasite' on the other. Thus, the model appears to simulate the 'mechanics' of parasitism embodied in the instability of the utility maximizing equilibrium.

The stable case is no less interesting. In that case, the two individuals settle down on an 'optimum' rate of production. However, this optimum is only a local maximum of utilities. The 'satisfaction' of both attains a larger maximal value elsewhere in the space of production rates. That 'really' best state, however, is not an equilibrium and cannot persist if each attempts to maximize *his own* utility. Only if the utilities are linked by a dependence or, alternately, if the production rates are interdependent, can the absolute maximum be reached.

Later I learned that the mathematical apparatus used in this model had been developed a century earlier by Cournot [7] in the context of a duopoly theory. And it was at the Center for Advanced Study in the Behavioral Sciences, that I saw the obvious connection between the basic idea of the model and that of the non-constant-sum game, in particular Prisoner's Dilemma. For in that game, too, attempts on the part of each player to maximize his own utility lead to an equilibrium, but not to an optimal outcome, whereas, the optimal outcome is not an equilibrium.

I referred to this connection in two papers published in 1956 [8] still dealing with 'parasitism' and 'symbiosis'. At that time I began to toy with the idea of putting the model to experimental test. The two individuals were to regulate their outputs by adjusting knobs. An analogue computer was to determine the resulting utilities projected on an oscilloscope, which would guide the subjects' attempts to maximize utilities. The question to be answered was whether in the case of the stable equilibrium, the subjects would be 'trapped' in it, unaware that by coordinating their adjustments they could both get larger utilities at the 'Pareto-optimal' point; or whether they would eventually learn to establish a tacit coordination and so reach the mutually beneficial rate of production.

It soon became apparent, however, that instrumentation problems would postpone the actual experiments indefinitely, and the project was abandoned. At the same time, it dawned on me that the Prisoner's Dilemma game captures the essentials of the idea embodied in the parasitism-symbiosis model and that experiments with it require practically no apparatus and moreover yield unambiguously quantifiable results in terms of frequencies of choices of the one or the other strategy.

However, being involved in other research, I still did not undertake 'straight' systematic experiments with Prisoner's Dilemma. Instead I incorporated a three-person version of the game in some experiments on the effects of sleeplessness being conducted at the Mental Health Research Institute at that time. A three-person version of the game was included as one of four tests administered during 32 hours of sleep deprivation, each test lasting eight hours. By changing the order of the tests administered to different groups of subjects, differences, if any, attributable to different periods of sleeplessness could be assessed. All tests involved three-person groups of subjects. In this way, the format of the experiment dictated the use of the three-person game [9].

The question asked was whether prolonged sleeplessness would have an effect on the ability of the three players to arrive at a 'tacit collusion'. which leads to the cooperative solution of Prisoner's Dilemma. Since the game was played for almost eight hours (1200 trials), one might expect that at least in some cases such tacit collusion would be established. Our inexperience, however, introduced another factor. For humanitarian reasons, the subjects were allowed to take a coffee break after four hours of pushing buttons (to indicate their choices in the three-way game). To avoid *explicit* agreement on the cooperative solution, they were not allowed to discuss the game during the break, and they were monitored to insure compliance. Nevertheless, the change of performance across the break was dramatic. In the first four hour session hardly any collusions were achieved, which in retrospect seems hardly surprising. In variants of the three-person Prisoner's Dilemma where the *single* defector gets the largest payoff and the single cooperator suffers the most severe punishment, it is all but impossible for all three to 'balance' themselves on the precarious cooperative solution. Following the coffee break, however, collusion was achieved by a large majority of the groups. Evidently just taking coffee was enough to establish cooperation (by telepathy?). It did not occur to us to give the subjects coffee at separate tables in a control experiment.

In the article cited above, Schelling called attention to the importance of 'tacit cooperation between opponents'. The thrust of his argument was the recognition that the most important human conflicts are modeled more properly by non-zero-sum rather than by zero-sum games, inasmuch as the interests of the parties are seldom diametrically opposed. Even in war, the enemies are often forced by circumstances to 'cooperate' in the sense of exercising certain self-imposed constraints in order to avoid escalation harmful to both. Schelling went on to spell out the nature of these constraints and the role of (often tacit) communication in strategic conflict. None of these ideas has any relevance to the theory of the zero-sum game, because in that context 'cooperation' between the players makes no sense: there is no outcome that is preferred by *both* to another outcome. Next, opportunity to communicate has no bearing on the theory of the zero-sum game. The outcome of a game with a saddle point, played by rational players is determined; hence it is neither advantageous nor disadvantageous to reveal one's strategy to the opponent (if he is rational). In a zero-sum game without saddle points, where mixed strategies are optimal, it is always a disadvantage to communicate to the opponent the particular strategy one intends to use on a given play of the game. In contrast, in non-zero-sum games, it is sometimes of great advantage to communicate one's strategy choice to the other (provided the communication is credible); for this information may force the other to use a strategy that confers an advantage on the first player. The game of Chicken is a famous example of such a situation. If one player can make the other believe that he is irrevocably committed to the 'daring strategy', the other has no choice but to 'chicken out' if he wants to avoid disaster.

Subsequently, Schelling and those who followed his lead, notably Jervis [10], made these aspects of 'strategic communication' central in the analysis of international relations. While global politics was definitely recognized as a 'non-constant-sum game', the central problem in the analysis remained implicitly that of finding 'optimal strategies', 'optimal policies', or the like. The object of the analysis was to provide insights into the dynamics of bargaining (threats, promises, commitments, bluffs) in the light of which the designers of policies and decision makers could achieve a degree of 'virtuousity' that would put them ahead in the game.

In reading Schelling's article, I too was impressed with the sterility of the zero-sum game model of serious human conflicts. However, I saw in his discussion of non-zero-sum models an attempt to introduce new techniques for the attainment of the same goals that are inherent in the zero-sum model: maximization of one's own utility. To be sure, the non-zero-sum model reveals that attempts to maximize one's own utility need not be identical with attempts to minimize the other's utility (as in zero-sum games). But the focus of attention was still on the problem of finding 'optimal strategies' from the point of view of individual rationality. In my opinion, this stance obscures the principal lesson to be drawn from the analysis of non-constant-sum games, namely that in those games individual and collective rationality are often at cross purposes and that the beneficial effects of the latter can be attained in only two ways: (1) by changing non-cooperative games to cooperative ones where explicit agreements can always be enforced or (2) by abandoning individual rationality in favor of collective rationality. Prisoner's Dilemma became for me the paradigm illustrating either of these imperatives. I began to write profusely on the subject [11] and in 1962, together with Chammah undertook systematic experimental investigations of the game, which were published in book form in 1965 [12].

As pointed out, other investigations began still earlier but the method really 'took off' in 1965 or thereabouts. Figure 1 suggests that the peak may have been reached around 1967, and it remains to be seen whether the 'explosion' peters out or becomes stabilized at some level determined by the availability of publication outlets.

Doubtless it is the 'conflict-cooperation' or 'cooperation-competition' dichotomy of Prisoner's Dilemma that attracts the experimenting social scientist, who sees in this simple 2×2 game an ideal experimental instrument. It offers the opportunity to gather large masses of data at small cost in time and money. The data are naturally quantifiable without resort to scaling techniques. Moreover, behavior in Prisoner's Dilemma can be spontaneous (in rapid repetitions) thus revealing the underlying motivational structure, which may often be masked in attempts to elicit it by verbal responses. The question remains, to be sure, to what extent the laboratory findings shed light on the dynamics of 'real life' conflicts. This question is at the center of lively controversies, and we shall return to it. For the moment let us see what has been done in the last six or seven years of massive experimentation with Prisoner's Dilemma and closely related games.

The traditional laboratory experiment is designed to answer specific questions posed as hypotheses. These involve some relations among two sets of variables: independent variables, manipulated by the experimenter, and dependent variables, whose variations in consequence of changing the independent variables answer the questions posed. The values of the dependent variables are read off from processed data. The most frequently examined dependent variable in Prisoner's Dilemma experiments is the relative frequency of cooperative choices (usually designated by C) in a group of subjects. This frequency provides a natural measure of the 'amount of cooperation'. There are other dependent variables of equal, perhaps greater interest, for instance the relative frequencies of all the four outcomes and the conditional frequencies of C choices following each of these four outcomes. These conditional frequencies suggest intriguing psychological interpretations. For instance the frequency with which a subject chooses C (i.e., 'cooperates') just after a double cooperative outcome (CC) has occurred suggests a measure of 'trustworthiness': he does not take advantage of the other's willingness to cooperate by switching to the immediately rewarding D strategy. On the other hand, the tendency to repeat C after one has just cooperated *without* reciprocation suggests a determination to 'teach by example', to try to induce the other to cooperate even at the cost of receiving the worst of the four payoffs. Other conditional frequencies can be interpreted accordingly. Clearly, these conditional tendencies to choose C are refinements of the over-all crude tendency, hence are all 'measures of cooperation'.

The independent variables used most frequently in Prisoner's Dilemma experiments fall roughly into four categories.

(1) *Payoffs*. The crux of the dilemma is that D (the 'defecting choice') dominates C, hence is dicated by individual rationality. The outcome CC, however, jointly dominates DD, hence is prescribed by collective rationality. The dominance of D over C has two components, namely 'greed' – the hope to get the largest payoff if the other tries to cooperate, and 'fear' – the tendency to avoid the unilateral cooperator's 'sucker's payoff'. By changing the relative magnitudes of the eight payoffs of the game, the relative importance of the cross pressures (to cooperate and to 'defect') can be assessed.

(2) *The Strategy of the Other*. One of the players can be a confederate of the experimenter, playing a pre-programmed strategy in iterated plays of the game. By comparing the real subject's behavior in response to various programmed strategies, the experimenter may hope to learn which strategies elicit the 'most cooperation', for example, the best 'mix' of C and D choices or the best mix of rewarding the subject's cooperative responses and punishing the defecting ones.

(3) *Information and Communication*. The game-theoretic paradigm assumes complete information about the structure of the game. On the other hand, the paradigm of the non-cooperative game precludes any communication between the players except through their choices revealed after they are made in iterated play. In the experimental setting, however, both information and communication conditions can be varied. Information about the structure of the game (for example about the other's payoffs) can be partly withheld; limited communication (e.g., occasional standard messages) can be allowed.

(4) *The Subjects Themselves*. These can be assessed on 'personality scales' or chosen from different population, characterized by sex, age, background, nationality, etc.

Besides relating behavior to these manipulable variables, performance can be also examined in the time dimension. The question here is what, if anything, the subjects 'learn' in the course of iterated play.

The same question can be asked, of course, of the experimental community: what, if anything, has been learned from Prisoner's Dilemma experiments? It stands to reason that in any experiment on human behavior strikingly consistent results are not to be expected. Still some over-all impressions arise. With regard to payoffs, the results are the most straightforward. Increasing the payoff in any of the cells of the game matrix leads to an increase in the frequency with which the strategy containing that cell is chosen. In itself, the result is devoid of interest, being expected on common sense grounds. However it is instructive to see that 'in the mass', human behavior, at least in the laboratory, is predictable. Besides, it is of interest to assess the quantitative changes in choice frequencies brought about by changes in payoffs and through them the relative importance of the cross pressures. Thus it appears that 'greed' is a more effective instigator of defections than 'fear' and that the rewards associated with the double cooperative outcome are somewhat more effective in eliciting cooperation than punishments associated with the double defective outcome. With regard to information and communication, the results are again in the expected direction. In general, both information and communication are conducive to the establishment of cooperation. Again, however, it is the quantitative rather than qualitative results that are of interest: how effective are different kinds of information or communication? It has been established, for example, that information irrelevant from the game-theoretic point of view is nevertheless of psychological relevance, such as a preliminary meeting, even though entirely cursory, with the co-player.

The effects of the strategy of other are almost as clear cut as those of the payoffs. By and large, the 'amount of cooperation' in the programmed player's strategy has little or no effect on the subject's behavior if the strategy is non-contingent; that is, if the other's responses are prescribed without reference to the subject's responses. Notable exceptions are the 'pure' cooperative or non-cooperative strategies. The totally non-cooperative strategy elicits, as expected, very low levels of cooperation. The totally cooperative strategy, on the other hand, typically elicits either high cooperation or, on the contrary, 'exploitation', thus apparently separating subjects into 'cooperators' or 'exploiters'. As for the contingent strategies, the most effective one in eliciting the subject's cooperation appears to be Tit-for-tat, where the programmed player matches the subject's previous choice. Perhaps it is the most effective because it is the easiest of the contingent strategies to discern. The best strategy against Tit-for-tat is the choice of C throughout except on the last play, after which no retaliation can follow.

The personality of the subjects is the least satisfactory of the independent variables in that it is most difficult to quantify and in that it yields the most ambivalent results. It is however the variable of greatest interest to those psychologists who view behavior in Prisoner's Dilemma as a sort of projective test and hope to find significant correlations between it and the motivations supposedly induced by the personality profiles of the players. Some positive results have been reported in this area, for instance by Terhune [13] who noted quite distinct patterns of behavior in three groups of subjects characterized respectively by high 'achievement need', high 'affiliation need' and high 'power need'. Also subjects recruited from different overtly recognizable populations (sex, background) behave differently at least in very long iterated runs of Prisoner's Dilemma.

Turning to the time courses of choices in iterated plays, we find an interesting typical pattern. On the whole, initial choices of C and D are about equally frequent. In the early phases of a long protracted run, the frequency of C choices tends to decrease, as if the subjects were learning that 'cooperation' does not pay in this game, learning, that is, to choose in accordance with 'individual rationality'. Eventually, however, the frequency of cooperative choices and especially of double cooperative choices increases, as if the subjects were finally learning to act in accordance with collective rationality. This increase, however, reflects only the 'average behavior' in massed protocols. Examination of separate protocols reveals that while some subject pairs learn to cooperate, others become entrapped in the DD outcome. The finding is a rough corroboration of the fundamental instability of the situation. It reminds strongly of Lewis F. Richardson's model of the unstable arms race, where an equilibrium is theoretically possible but where, in effect, the arms race can only either escalate indefinitely or, be reversed, leading to disarmament and steadily increasing cooperation.

Having examined the over-all results of extensive experiments with Prisoner's Dilemma as a laboratory simulation of typically ambivalent human conflict, we can try to answer the question of what has been learned and, implicitly, what is the value of experiments of this sort. In particular, do the results teach us anything about the dynamics of conflicts or techniques of conflict resolution? If the demand is for concrete practical 'know-how', the answer is clearly no. It is not merely a matter of rejecting foolhardy generalizations from laboratory to life. (Who would care to infer from the effectiveness of Tit-for-tat the advisability of guiding one's life by the eye-for-an-eye principle?) The difficulty is to find agencies both willing and able to learn from the one important lesson inherent in the Prisoner's Dilemma model, namely that 'individual rationality' can in many conflict situations be a trap. This lesson emerges from the logical analysis of the game. Whatever is observed in the laboratory simply calls our attention to the different conditions under which players fall or do not fall into that trap and how they sometimes get out of it when they do. These demonstrations are of no consequence to some one who by virtue of his position as policy maker or decision maker must orient himself to the problem of finding 'optimal strategies'. On the other hand some one who wants to guide his life by collective rationality as a matter of principle (the Kantian imperative), will interpret the results of the experiments as simply confirmations of his view. He may be encouraged by the finding that among the individuals called upon to play Prisoner's Dilemma just once (when the cooperative choice cannot be rationalized by individual rationality), about 40% choose C; or he may be discouraged by the finding that about 60% choose D. But a commitment to collective rationality is not likely to be affected by this finding, because the commitment is one of principle, not one based on 'maximizing expected payoff'.

Thus, the value of the experimentation is to be sought not in specific findings but in the arousal and dissemination of interest in the idea. So far, the Prisoner's Dilemma 'fad' has had two consequences in research circles, both being technical developments in the theory of the non-cooperative game. Recall that the 'paradox' of the game calls in question the significance of the equilibrium outcome as a 'solution' of all such games. The identification of the 'solution' with an equilibrium is a straightforward generalization of the 'solution' of the constant-sum game. There the rationality of the equilibrium cannot be challenged. However, the compelling nature of this solution rests on deeper grounds. It is shown that if a two-person constant-sum game has several equilibria, they are all 'equivalent' and 'interchangeable'. They are equivalent in the sense that the payoffs to each of the two players' are equal at all the equilibrium outcomes. They are interchangeable in the sense that if each player chooses *any* of the strategies containing an equilibrium, the outcome is an equilibrium. Therefore in the context of the two-person constant-sum game it is possible to make an unambiguous prescription to each player: Choose a strategy that contains an equilibrium. The result will be a defensible 'solution' of the game.

In analyzing the general (non-constant-sum) non-cooperative game, it has been shown that every such game contains at least one equilibrium. However, if there are several equilibria, they need be neither equivalent nor interchangeable. This makes it impossible to prescribe an optimal strategy to the individual players even if one assumes that any equilibrium outcome is a defensible 'solution' of the game. This is because, if the equilibria are not interchangeable, such a prescription, if followed may result in an outcome that is not an equilibrium, hence not defensible as a solution.

Harsanyi [14] holds it as axiomatic that 'rationality' of both (or all) players implies a unique defensible solution of a non-cooperative game. (Recall that the concept of rationality includes a conviction that both or all players are rational like oneself.) Accordingly, Harsanyi undertook to remove the ambiguity introduced by non-equivalent and non-interchangeable equilibria of the general non-constant-sum game. To do this he, in effect, introduces the notion of 'tacit bargaining', a bargaining that would take place if communication were allowed. If also enforceable agreements were possible, communication, whether explicit or tacit, would turn the non-cooperative game into a cooperative game. This is, however, where Harsanyi draws the line. He distinguishes the non-cooperative game not by the circumstance that communication is excluded from it (since it is replaced by 'tacit' communication) but by the exclusion of *enforceable* agreements. In the absence of such agreements only equilibria can be considered as possible solutions. However Harsanyi's bargaining procedure enables him to remove from considerations all equilibria but one (or a set of equivalent ones).

Now Prisoner's Dilemma has a single equilibrium (DD), which is trivially equivalent to and interchangeable with itself. Therefore in Harsanyi's scheme it is *the* solution of the game. Had game theory remained on the formal level, there would be nothing further to say about this game. However, because the game aroused so much controversy, the 'solution' that violates collective rationality has to be defended. Harsanyi's defense is most interesting. It does not issue as a defense of 'toughmindedness' versus 'idealism' of the sort one often encounters among the adherents of the so called 'realist' school of international relations. Rather it is an argument about the *inevitability* of the 'bad' solution in the absence of enforceable agreements and can thus be construed as an argument in favor of enforceable agreements if the 'bad' equilibrium is to be avoided. The argument does not specify the nature of the 'enforcements'. One can, therefore if one wishes, (at least this is my interpretation) invoke 'internal' enforcements based on self-imposed (conscience determined) sanctions for breaking the agreement. From the moral point of view, therefore, Harsanyi's defense of the non-cooperative 'solution' of Prisoner's Dilemma is by no means reprehensible. It need not be construed as an 'advice' to players to play noncooperatively, but on the contrary, as an advice to seek ways of effecting enforceable agreements so as to turn the uncooperative game into a cooperative one. If my interpretation is correct, then Harsanyi's view does not differ essentially from my own.

Another development along similar lines was undertaken by Howard [15]. Howard introduces the notion of 'meta-strategies'. A meta-strategy is to an ordinary strategy what a move is to a strategy in an ordinary game. Thus one player's meta-strategy in Prisoner's Dilemma would be a decision of which of the two strategies to choose contingent on the other's choice, *if that choice were known*. Thus in Prisoner's Dilemma each player has four meta-strategies of the first order (four mappings of one's own two strategies upon the two strategies of the other). However a choice of a meta-strategy of the first order by each player may not determine the game. Another step is required to meta-strategies of the second order, of which each player has sixteen (the mappings of the four meta-strategies of the first order on the other's two simple strategies). The meta-games examined are those in which one of the players has four first order meta-strategies and the other sixteen meta-strategies of the second order. In general, outcomes that are not equilibria in the original game may turn out to be equilibria in the meta-game. Moreover, further extensions to higher order meta-games yield no new equilibria in two-person games. The interesting result is that in the meta-game derived from Prisoner's Dilemma, the cooperative (CC) outcome turns out to be an equilibrium (in addition to the original equilibrium) and so a defensible solution of the game in terms of 'meta-rationality', which in this case coincides with collective rationality. Howard's procedure like Harsanyi's also introduces 'tacit bargaining'. However the results transcend Harsanyi's in the sense that Pareto-optimal outcomes that may not have been equilibria in the original game turn out to be equilibria. Thus the 'paradox' of Prisoner's Dilemma can be resolved by invoking a 'higher' strategy space. Such resolutions have been observed in the history of mathematics when apparent 'paradoxes', (e.g., incommensurability of well defined quantities) were resolved by extending the conceptual reportoire (e.g., extending the concept of number from rationals to reals).

Finally, preoccupation with Prisoner's Dilemma as 'the' standard representative of the non-constant-sum non-cooperative game led to examination of other such games. In 1966, Guyer and I published a list of 78 types of 2×2 games (involving two players with two strategies each) [16]. The discrete classification was made possible by restricting the payoffs to an ordinal scale so that only the ordinal magnitudes of each player's four payoffs, as they are placed in the game matrix, determines the 'species' of the game. These 'species' were then grouped into 'genera', 'orders', and 'classes' depending on the pressures operating on one or both players to choose one strategy in preference to the other. The richness of the variety revealed the formidable complexity of even the simplest of strategic conflicts, once one considers not only the 'rational solution', whichever way one chooses to define 'rationality', but also all the cross pressures that introduce ambivalence into most of these situations. The majority of these games turn out to be asymmetric. That is, the strategic positions of the two players are unequal in them. They thus provide the opportunity of studying the behavior of the 'underdog' vis-à-vis the 'topdog' in situations with a skewed distribution of 'power'. Subsequently several investigators undertook experimental work on other than Prisoner's Dilemma games.

This latter development may, perhaps save the current interest in experimental games from extinction. If Prisoner's Dilemma has not yet been mined dry experimentally, there is probably sufficient payoff in the form of interesting, possibly surprising findings to be extracted from the other 77 varieties (or at least a dozen or so distinctive categories) of 2×2 games to keep a generation of investigators busy.

In summary, the impact of Tucker's anecdote about the two prisoners has been on the perception of human conflict as something different from a problem of finding a 'utility maximizing course of action', which is the principal theme of classical decision theory. The value of this change of perception depends on how pervasive it becomes. In circles where policies and far-reaching decisions affecting the entire human race are made, 'rationality' is still predominantly identified with strategic calculations be they 'cost benefit analyses' or 'image building' with the view of gaining for one self or for one's client the most advantageous attainable position in competition for resources, influence or power. This is at best 'zero-sum mentality' – at best if it takes into account the countervailing efforts of opponents and attributes rationality to them. Usually calculations of this sort are not even cast in the form of zero-sum games but rather implicitly as games against nature, where nature is modeled as a stochastic environment and the only 'active' decisions are made by the actor whose interests are to be promoted.

The zero-sum model is one step removed from the game against nature: it takes into account the countervailing efforts of an opponent and attributes rationality to him.

Both the game against nature model and the two-person zero-sum game model are extremely seductive, because the problems posed by them can in principle be solved unambiguously and moreover often challenge the ingenuity of the analyst. Consider the complexity of a 'game' between an ICBM, programed with evasive strategies and an ABM programmed with pursuit strategies calculated to frustrate the evasive strategies. Solution of such problems brings understandable satisfaction to the scientific entourage of military establishments. A perusal of the literature on applications of game theory to logistic and military problems reveals that *all* of them are cast in the form of two-person zero-sum games. But 'optimal' solutions of such games are obviously available to both sides and the net result of the expenditure of intellectual effort (not to speak of resources) is simply a growing sophistication of the 'art of war'. Benefits of this development accrue at most to the practitioners of the profession. They are a threat to the rest of humanity. For this reason, deflection of attention at least in intellectual circles away from the 'classical' problems of decision theory, especially from its augmentation by the theory of the zero-sum game, and the consequent raising of searching questions about the meaning of 'rationality' in situations other than games against nature and conflicts with diametrically opposed interests is to be welcomed.

Dept. of Psychology, Univ. of Toronto

BIBLIOGRAPHY

[1] Flood, M. M., 'Some Experimental Games', Research Memorandum RM-789, The Rand Corporation, Santa Monica, 1952.
[2] Luce, R. D. and Raiffa, H., *Games and Decisions*, John Wiley and Sons, New York, 1957.
[3] Schelling, T. C., 'The Strategy of Conflict: Prospectus for the Reorientation of Game Theory', *Journal of Conflict Resolution* 2 (1958) 203–264.
[4] Deutsch, M., 'Trust and Suspicion', *Journal of Conflict Resolution* 2 (1958) 267–279.
[5] Guyer, M. and Perkel, B., 'Experimental Games: A Bibliography 1945–1971', *Communication* No. 293, Mental Health Research Institute, The University of Michigan, March, 1972: bibliography in *Cooperation and Competition: Readings on Mixed Motive Games* (ed. by L. S. Wrigtsman, Jr., J. O'Connor, and N. J. Baker), Brooks/Cole Publishing Co., Belmont, Calif., 1972; Shubik, M., Brewer, G., and Savage, E., 'The Literature of Gaming, Simulation and Model Building: Index and Critical Abstracts', Report R-620-ARPA, Rand Corporation, Santa Monica, 1972.
[6] Rapoport, A., 'Mathematical Theory of Motivation Interaction of Two Individuals', *Bulletin of Mathematical Biophysics* 9 (1947) 17–27.
[7] Cournot, A. A., *Recherches sur les principes mathématiques de la théorie des richesse*, Libraire Hachette, Paris, 1838.
[8] Rapoport, A., 'Some Game Theoretical Aspects of Parasitism and Symbiosis', *Bulletin of Mathematical Biophysics* 18 (1956) 15–30.
Rapoport, A. and Foster, C., 'Parasitism and Symbiosis in an n-Person Non-Constant-Sum Continuous Game', *Bulletin of Mathematical Biophysics* 18 (1956) 219–231.
[9] Rapoport, A., 'Some Self-organizing Parameters in Three-Person Groups', in *Principles of Self-organization* (ed. by H. von Foerster and G. W. Zopf, Jr.), Pergamon Press, New York, 1962.
[10] Jervis, R., *The Logic of Images in International Relations*, Princeton University Press, Princeton, N.J., 1970.
[11] Rapoport, A., 'Critiques of Game Theory', *Behavioral Science* 4 (1959) 49–73; *Fights, Games, and Debates*, University of Michigan Press, Ann Arbor, 1960.
[12] Rapoport, A. and Chammah, A. M., *Prisoner's Dilemma*, University of Michigan Press, Ann Arbor, 1965.
[13] Terhune, K., 'Motives, Situation, and Interpersonal Conflict within Prisoner's Dilemma', *Journal of Experimental and Social Psychology* 6 (1970) 187–204.
[14] Harsanyi, J., 'Rationality Postulates for Bargaining Solutions in Cooperative and Non-Cooperative Games', *Management Science* 9 (1962) 141–153.
[15] Howard, N., *Paradoxes of Rationality*, The MIT Press, 1971.
[16] Rapoport, A. and Guyer, M., 'A Taxonomy of 2×2 Games', *General Systems* 11 (1966) 203–214.

STRUCTURAL PROPERTIES AND RESOLUTIONS OF THE PRISONERS' DILEMMA GAME*

1. Introduction

In this paper we make use of a theory of social behavior (Burns and Meeker, 1973; 1972) to analyze structurally the prisoners' dilemma game (Rapoport, 1960; Rapoport and Chammah, 1965) and to indicate resolutions of the dilemma under specified conditions. The theory is characterized by (i) descriptive models rather than normative or prescriptive models of social behavior; (ii) emphasis on multi-dimensional processes and on structural relationships rather than uni-dimensional quantities (such as 'utility'); and (iii) a view of interaction processes and context as a system (Buckley, 1967) and, in particular, the thesis that evaluation, decision, and interaction processes cannot be understood apart from the social context in which they occur (Burns, 1973).

In the theory, a social situation in which a number of actors are engaged is conceptualized as an *interaction system*. It consists of the following:

(i) *The set of m-actors and for each actor an action structure*: a set of relevant action alternatives, perceived outcomes of acts, preference structure with respect to outcomes, and a decision procedure.

(ii) *Meta-processes*, that is higher order processes, selecting or determining, among other things, the preference structures, relevant action possibilities, and decision procedures of the actors.

(iii) *Constraints* which determine actors' action and interaction opportunities in the situation: for example, permitting or not permitting communication and binding agreements (in game theory terms, whether a 'cooperative' or 'non-cooperative' game obtains).

In our formulation, actors may relate to one another on several levels. We are interested not only in their transactions on the *behavioral level* (the level with which classical game theory is concerned), but in interactions at meta-process levels; for example, interpersonal influence on one another's preference structures or decision-making. At such levels, actors' established relationships – friendship, kinship, etc. – are of special significance (Burns, 1973). In the analysis of social interaction, we emphasize the influence of existing social relationships among the actors on their preference structures, choices, and interaction patterns. The ultimate object of our formulation is to describe and to predict preferences, decisions, and interaction patterns, and in this particular paper, those likely to occur in the prisoners' dilemma game under various social conditions.

The paper is divided into three parts. In Section 2, we introduce our basic concepts and theoretical models. In Section 3, we structurally characterize and analyze the prisoners' dilemma game. Section 4 considers resolutions of the game, as indicated by our theory.

2. Basic Concepts and Theoretical Model: Interaction Systems[1]

A. *Basic Action Concepts*

The major concepts of the theory are outlined below.

i. *Actors.* A finite set of m actors $\mathcal{M} = \{1, 2, ..., i, ..., j, ..., m\}$. An actor is a person or group of persons with concordant interests.[2] In the purest case, there are no conflicting interests among the members of the group.

ii. *Actions.* Each actor has a finite set of possible action alternatives of which, in any given situation, a finite subset A^i (for actor i) are relevant or considered for choice.

iii. *Action outcomes.* Associated with each act α in the set A^i of action alternatives considered by actor i is a finite subset $H^i(\alpha)$ of *perceived outcomes*. Each outcome is assumed to be representable as a point x in a fixed n-dimensional space $\mathbb{R}^n$.[3] The set of all perceived outcomes (for all actors) is denoted by $\Theta = \{x^1, x^2, ..., x^n\}$.

iv. *Action-outcome structure.* This is the family of mappings H^i of A^i ($i = 1, 2, ..., m$) into the set of all subsets of all outcomes, 2^Θ. As indicated above, $H^i(\alpha)$ is the set of possible outcomes for action α in the given situation.[4]

v. *Preference structures.* A preference structure is an order relation on a set of items. This paper focuses on preference structures on action outcomes (R) and preference structures on action alternatives (Q). We assume all preference structures to be irreflexive relations and hence representable as a digraph (see Appendix and Burns and Meeker, 1973).

Preference structures are not necessarily complete orderings.[5] Non-comparability, multi-dimensional evaluation, and related properties such as intransitivity, inconsistency, and indecidability are basic features of the theory.

Actors are assumed to evaluate entities from different frames of reference. Evaluation of a good or activity on the basis of tradition may differ from evaluation of its use or intrinsic value to the evaluator or to others about whom he is concerned. The result is a variety of preference structures with different underlying bases. (We have formulated methods for the comparison and classification of order relations (see Appendix and Burns and Meeker, 1973) and a model of the process (meta-process) of evaluating and choosing among alternative preference relations.)

vi. *Decision procedures.* Decision procedures (d) such as 'maximization', 'max-minimization', 'satisficing', etc. are used to construct preference structures over action alternatives from preference structures over outcomes.

vii. *Indecisiveness.* As mentioned above, preference structures are not assumed to be complete (there may exist non-comparable elements) nor even transitive, and hence do not necessarily provide a partial order. Empirical orders will often include inconsistencies (because of the multi-dimensional nature of the outcomes and evaluations of them) and will only rarely have a unique maximal or most preferred element. A major concern of our work is how an actor responds, given the inconsistencies and contradictions inherent in most preference structures, to particular situations in which he finds himself.

viii. *Meta-processes.* The selection of a preference structure or of a decision procedure, the modification of a preference structure so as to remove inconsistencies or to decrease indecisiveness, and the search for new action alternatives or outcomes are examples of meta-processes used by actors in response to a particular context, to other actors and social pressures, or to the lack of a most preferred alternative.

ix. *Choice behavior.* Choice behavior consists of selecting an alternative or subset of alternatives from among a set of alternatives. The bases of choice are several: selection of the most preferred alternative evaluated by the actor on the basis of his goals and values, selection prescribed by authority or tradition, or random selection. The point is that there exist various bases whereby actors choose an action or subset of actions from among the set. It is not our intention here to exhaustively discuss the various choice bases, but merely to emphasize the existence of their multiplicity. In this paper, the focus is on the construction of preferences over action alternatives from preferences on outcomes rather than on orderings based, for example, on authority or tradition (see Burns and Cooper, 1971).

Given a choice from among a set of alternatives over which there is a preference structure, the actor chooses the most preferred alternative provided one exists. In the absence of an unequivocal maximal element, the actor strives, by various meta-processes, to generate such a preferred alternative.

x. *Social orientation and interaction.* The preceding outlines the process of individual multi-dimensional evaluation and decision-making. In a social situation, the outcomes which each actor experiences depend not only on his own behavior but also on that of others whom, in most instances, he cannot completely control. Social interaction situations are characterized by multiple action vectors $\alpha = (\alpha_1, \alpha_2, ..., \alpha_m)$ for a set of m actors $\mathcal{M}$ and the set of possible outcomes Θ. The outcome(s) associated with the action vector α are those of the set $\Theta(\alpha) = \bigcap H^i(\alpha_i)$ (see page 41 and Burns and Meeker, 1973).

Outcome evaluation and choice behavior may depend not only on an actor's own personal preferences but on those of other actors with whom he interacts or has social relationships. Actors' *social attitudes or orientations* toward one another reflect the extent and manner they take one another into account in their behavior. Social orientations with respect to self, another actor, a class of actors, or a collectivity are meta-processes which act upon preference structures and decision procedures of the actors involved, yielding re-evaluations (new or different preference structures) and/or changes in decision procedures.

B. *Action Structure*

We suppose that actor i has a preference structure R^i over the set of perceived outcomes.[6] R^i is considered as a subset of $\Theta \times \Theta$ in the usual fashion[7] and is assumed to be an irreflexive relation (all pairs (x, x) are excluded from R^i) and therefore representable as a digraph (see Appendix and Burns and Meeker, 1973, 1972). On the basis of the perceived outcomes, the preference structure R^i and the outcome sets $H^i(t)$, $t \in A^i$, actor i will seek to form a ranking (preference structure) over his viable action alternatives A^i. This problem can be modeled by the 4-tuple $E^i = = (\Theta, R^i, A^i, H^i)$ which we refer to as an *action-outcome-preference structure* or *AOP-structure*.

An actor constructs (or obtains) a preference relation over action alternatives from AOP-structures by means of *decision procedures*. For example, in the case of a one-dimensional utility function (a map of Θ into $\mathbb{R}$) one thinks of the 'maxmin' and 'maxmax' procedures of decision theory (see below). The selection, by i, of a given procedure d^i to apply to his AOP-structure E^i results in a (possibly trivial) preference relation or ordering over the set A^i of viable action alternatives, $Q^i = Q(E^i, d^i)$. The 5-tuple $C^i = (E^i, d^i) = (\Theta, R^i, A^i, H^i, d^i)$ is called an individual *action structure*.

In most instances the contradictions, inconsistencies, and incomparabilities of the preference relation, R, do not, for any given decision procedure, permit the establishment of an unequivocally preferred action (that is, Q^i, does not define a unique maximal element α_i^* in A^i). For this reason, the relationship between the AOP-structure and various decision procedures yielding order relations on A^i becomes of primary interest. Also brought to prominence are the meta-processes by which actors attempt to resolve decision problems in the midst of conflicting priorities.

Let us focus on an individual actor i and his AOP-structure, $E^i = = (\Theta, R^i, A^i, H^i)$. We omit the superscript 'i' for convenience. We outline several common decision procedures generalized in terms of our formulation.

(1) *Optimizing* (weak form). In this case, the actor searches for the

alternative whose outcomes dominate those of all other possible actions. This procedure can be modeled here by defining, for α and β in A, $(\alpha, \beta) \in Q_{OP}$ if, for some $y \in H(\beta)$ there exists an $x \in H(\alpha)$ with $(x, y) \in R$ and, for all $y \in H(\beta)$, $z \in H(\alpha)$, $(y, z) \in R$ implies $(z, y) \in R$ as well.[8]

(2) *Maximization and minimization.* A maxmax decision procedure focuses on only those elements of the set Θ of highest evaluation or rank (the outcomes in $L(R, O)$, see Appendix) and seeks to rank alternatives in terms of such elements in $H(\alpha)$ (the points of $H(\alpha) \cap L(R, O)$), $\alpha \in A$. Denote such elements by $H_M(\alpha)$. For a pair of actions α, β in A we write $(\alpha, \beta) \in Q_M$ if for each $y \in H_M(\beta)$, there exists an $x \in H_M(\alpha)$ with $(x, y) \in R$ and for all $z \in H_M(\alpha)$, $y \in H_M(\beta)$, $(y, z) \in R$ implies $(z, y) \in R$.

Maximizing procedures on the directional dual of R (the digraph obtained by reversing the pairs of R, i.e., $(x, y) \in R$, becomes (y, x) in the directional dual) becomes minimizing procedures for R.

(3) *Maxmin Procedures.* The common maxmin procedures can be formulated in the generality of our model by considering each of the sets $H(\alpha)$, $\alpha \in A$ as a subgraph of R. The *minimal elements* in $H(\alpha)$ are those in the strong components (of $H(\alpha)$) which, in the condensation digraph of $H(\alpha)$, have outdegree O (see Harary et al., 1966, p. 16, and the Appendix). Let us denote these elements by $H_m(\alpha)$. For a pair of actions α, β in A we write $(\alpha, \beta) \in Q_{Mm}$ if, for some $y \in H_m(\beta)$, there exists an $x \in H_m(\alpha)$ with $(x, y) \in R$ and for all $z \in H_m(\alpha)$, $y \in H_m(\beta)$, $(y, z) \in R$ implies $(z, y) \in R$.

There are many other order relations possible on the action alternatives. We speak of such procedures as involving *outcome analysis*. These are to be contrasted with decision procedures which entail little or no consideration of the actual outcomes. For example, societal norms may prescribe 'one does this but not that'. It may also happen that persons or groups in positions of authority prescribe actions to be taken in certain contexts or, perhaps, provide a scalar-valued utility function.

C. *m-Actor Action Structure*

In this section, we present a model capable of representing many types of social interaction (Burns and Meeker, 1973; 1972).

The individual AOP-structures of the various actors engaged in a social setting can be combined to form the *m-actor AOP-structure*,

$$E(\mathcal{M}) = (\mathcal{M}, E^1, E^2, ..., E^m).$$

This structure, with the addition of a decision procedure used by each actor to rank his action alternatives, becomes the *m-actor action structure*, $\Gamma(\mathcal{M}) = (\mathcal{M}, C^1, C^2, ..., C^m)$ (or if we wish to emphasize the decision procedures employed, we write $\Gamma(\mathcal{M}) = (E(\mathcal{M}), d^1, ..., d^m))$ where $\mathcal{M} = \{1, 2, ..., m\}$ denotes the finite set of actors involved in the situation and $E^i = (\Theta, R^i, A^i, H^i)$ is an AOP-structure and $C^i = (\Theta, R^i, A^i, H^i, d^i)$ an action structure for actor $i \in \mathcal{M}$.

If $\alpha_i \in A^i$, $i = 1, 2, ..., m$, then $\Theta(\alpha) = \bigcap_{i=1}^{m} H^i(\alpha_i)$, which we assume to be non-empty, is the set of possible outcomes resulting from the multiple action vector $\alpha = (\alpha_1, \alpha_2, ..., \alpha_m)$. We do not require that $\Theta(\alpha)$ contain but a single outcome. Should $\Theta(\alpha)$ be a singleton, $\{x\}$, for each α we shall say that the actors of $\mathcal{M}$ *control the situation*.

An actor i, with action α_i, satisfying $H^i(\alpha_i) = \{x\}$ has the possibility of dominating the situation if he so desires. If all the sets $H^i(\alpha_i)$, $\alpha_i \in A^i$, are small (few elements), i has a powerful position within the situation. This can be contrasted to the case of an actor whose alternatives yield large outcome sets, in the extreme, if $H^j(\alpha_j) = \Theta$ for all $\alpha_j \in A^j$, that is, j lacks any control over outcomes occurring and, therefore is *powerless* in the situation.

[It should be noted that the equation $H^j(\alpha_j) = \Theta$ also may in some cases represent j's alternative of 'no action'. Whether this represents a desired alternative, or not, depends upon the entire structure of the problem.]

Each actor's d^i results in an order relation Q^i on the set of i's alternatives A^i. In the event of complete decidability each pair (A^i, Q^i) defines a uniquely preferred alternative α_i^* resulting in the outcome set

$$\Theta(\alpha^*) = \bigcap_{i=1}^{m} H^i(\alpha_i).$$

In the general model formulated here, one cannot expect a 'solution to the game' as is commonly the objective in classical game theory. Indeed, as the preference relations R^i over the perceived outcomes Θ are not assumed to be complete (there may exist non-comparable outcomes) nor transitive, they do not in general yield even a partial order. An actor, who must make a decision under conditions where inconsistencies obtain, will often be forced to modify his action structure in such a fashion that the resulting preference relation Q over his alternatives indicates a preferred action. The manner in which this is done is strongly dependent upon the social relationships and influence processes among the various actors.

D. *Meta-Processes*

In addition to the m-actor action structure, there are meta-processes which characterize the interaction system.

The action structure for actor i, $C^i = (\Theta, R^i, A^i, H^i, d^i)$, can be considered a decision process resulting in social action. The processes of selection involved in constructing C^i are themselves decision processes, which we refer to as meta-processes.[9] For example, given an AOP-structure $E^i = (\Theta, R^i, A^i, d^i)$, an actor has many decision procedures available (maxmin, maxmax, satisfice, etc.), each resulting in a (usually) distinct action structure $C^i = (E^i, d^i)$. The process of selecting a decision procedure is an example of a meta-process. In this instance, we have a new higher order situation whose outcomes $\bar{\Theta}$ are action structures $(E^i, d^i)'$, whose preference relation $\bar{R}$ might be based on criteria such as simplicity of application, maximization of decidability, or some other *attribute of action structures*, and whose alternatives in this case are the decision procedures under consideration. One should note that the situation clearly exhibits other meta-processes, namely that of choosing the preference relation $\bar{R}$ and the decision procedure $\bar{d}$ to be used at this level.[10] This pyramiding of decision processes is an essential part of the model and, of course, entails a hierarchy of meta-processes. We can symbolize (as in Arrow, 1963, p. 90) the k-level system by the k-tuple $\{^1C^i, {}^2C^i, ..., {}^kC^i\}$ where $^{j+1}C^i$ is the process of choosing, let us say, a decision procedure for $^jC^i$. For any given system, this decomposition can be accomplished in many different ways depending upon the structure of primary interest.

In the analysis of the prisoners' dilemma situation, we are interested in how actors influence one another and how the context in which they find themselves affects them at the meta-process level so as to transform the interaction system into one of a different type. In particular, we consider how actors' social attitudes or orientations toward one another, as meta-processes, influence their preference structures and decision procedures and, therefore, their interaction patterns.

Let us consider the problem from the standpoint of actor 1. He perceives the set of outcomes Θ, his preference relation $R^1 \subset \Theta \times \Theta$, his action alternatives A^1, and a set of fellow actors $\mathcal{M}_1 \subset \mathcal{M}$. On the basis of his knowledge of actor j, $j \in \mathcal{M}_1$, he constructs a preference relation $R_1^j \subset \Theta \times \Theta$ representing his assessment of j's evaluation of the possible outcomes.[11] As a result of the interaction of the two preference structures R^1 and R_1^j, actor 1 may re-evaluate outcomes and action alternatives. In our analysis, an actor's preferences are investigated in the context of his beliefs and concerns about the preferences of others and of the preferences he would have if he held different beliefs (Shick, 1972).

If 1 is concerned primarily about j's welfare, referred to as *positive other orientation* ('altruism'), he looks for those outcomes ranked highest in R_1^j, that is, those contained in the left-hand side of $\sigma(R^1, R_1^j)$, the mutual level structure matrix (see Appendix and page 46). On the other hand, if a *negative other orientation* ('hostility') toward j prevails, 1 would focus on outcomes of low rank in R_1^j, that is, as near the right-hand side of $\sigma(R^1, R_1^j)$ as possible. In cases where 1 ignores (deliberately or unknowingly) actor j and concerns himself only with his own welfare, he would disregard R_1^j in his actual rankings, focusing on those outcomes ranked high in R^1, that is, in the upper part of the σ-matrix.[12] His choice of an acceptable alternative becomes that of the individual decision-maker discussed earlier. We refer to 1's orientations in this case as pure *self-orientation* ('individualism'). Were 1 to have dual concerns, regarding j as well as himself, referred to as *positive self/other orientation* ('cooperativeness'), he would look for outcomes preferred by both 1 and j, that is, the convergent or intersecting outcomes ranked highest in *both* R^1 and R_1^j, namely the entries of $\sigma(R^1, R_1^j)$ lying as close to the upper left hand corner as possible.

Social orientations between or among actors operate as meta-processes or higher order evaluation processes placing relative values on the various evaluations of outcomes (and outcome components). Thus, actors with pure self-orientations concern themselves only with their personal pay-offs while those with self/other concerns focus on outcomes preferred by both self and other, in the purest case giving equal value to these preferences. Still other orientations relate to rivalry or competitiveness (an actor valuing outcomes according to the degree his outcome component surpasses that of others, see Emshoff, 1970), equity and distributive justice, masochism, and so forth (Burns, 1972). Elsewhere (Burns, 1973) we have considered the social and institutional context associated with particular orientations.

E. *Constraints and the Availability of Action Opportunities*

The concept of 'constraint' can be counterpoised to that of choice (Burns and Cooper, 1971, p. 77–86). Behavioral constraints, above all those of a social and institutional nature, serve to determine actors' access to resources and action opportunities and, therefore, to define the limits within which actors make choices (the set A^i of viable action alternatives in a given context). In Buchler and Nutini's terms (1968, p. 7), these are the 'ground rules', for example, cultural norms and jural rules.[13]

The performance of most actions entail certain requisites: material and human resources, skills, knowledge, and other factors which are essential to executing the actions. Restrictions on an actor's access to these (e.g., they are under the control of other actors, including those with whom he interacts directly) may make it impossible for him to carry out certain acts, even when the action possibilities are known to him. In a similar manner, his performance of particular acts will be facilitated by his having access to or control over action requisites.

The constraints on i's and j's actions define the set of possible interactions between them (as represented by the matrix of interaction possibilities). Any of several factors may determine such bounds – different factors often producing the same effect – thereby constraining or facilitating i's and j's opportunities to engage in conflict or exchange generally, or in different types of conflict or exchange. This idea encompasses several types of factors among which are the following:

(i) Many action possibilities fail to exist simply because of the state of technology (in a cultural milieu or historical period) of the society in which the actors find themselves. Certain alternatives requiring particular tools or other mechanical means may simply not have appeared.

(ii) Many action alternatives exist 'objectively' but are unknown to the actor. In the limit, just a single action possibility may be perceived (e.g., that prescribed by tradition or by authority).

(iii) The actors may know of alternatives but remain unwilling or unable to consider them, lacking the time, capabilities, or opportunities to go through the process of collecting information, comparing, evaluating, and making decisions in regard to them (habitual and traditional forms of behavior are especially likely under such conditions).

(iv) Many actions are social. They require the cooperation or, at least the acquiescence of others (possibly, the other actors involved in the interaction). An actor's opportunity structure through which he has access to and control over action resources depends on his position in a power structure, a network of social relationships. His options may be strictly limited by the absence of rights, by the refusal of others to cooperate in the case of activities which require such cooperation, or by the normative constraints others place on the activities.

Persons with low social status or power, and therefore lacking control over essential resources (material and social) are unable generally to carry out many types of collective actions (either of an exchange or conflict nature). On the other hand, those with high social status and power (with access to or control over essential resources) have more action possibilities, including that of generating new action alternatives. Clearly, the social structuring of a society results in *differential constraints* on actors' possibilities of action ('opportunity structures').

Constraints on i's and j's action possibilities define the set of possible interactions between them, e.g., the possible types of exchange or conflict that can take place. In other words, the constraints on i and j serve to delimit or determine their viable action sets, A^i and A^j respectively, the constraints possibly affecting their action opportunities differentially. Thus, under one set of social and institutional conditions, we would expect an interaction situation to be represented by an interaction matrix consisting of the coupling of A^i and A^j. Under other structural conditions, a new matrix would obtain, defined by the repertoires $\bar{A}^i$ and $\bar{A}^j$ (where $A^i \neq \bar{A}^i$ and $A^j \neq \bar{A}^j$) determined by the situational constraints unique to those conditions. Through the actors' efforts as well as those of others, their repertoires may be changed, in some cases expanding, in others contracting (Burns and Meeker, 1972).

For instance, there may be constraints preventing the development of close or cooperative relationships between persons who come into contact with one another in their everyday work. Possession of certain social attributes such as those of age, sex, religion, tribe, or ethnic group may serve to dissociate some actors from (or on the other hand, associate them with) other persons or groups in the situation. Such conditions affect actors' access to important resources.

Action and interaction opportunities – and the constraints and facilities affecting these – are of more fundamental significance to behavioral analysis than 'choice behavior' as such, for they determine the limits of social interactions and, in particular, exchange and conflict: what actors can and cannot do, in particular, whether they can or cannot engage in mutual exchange or conflict, what types of exchange or conflict are possible, and so forth. Within the limits determined by actors' knowledge, performance capabilities, and social constraints – and only within these limits – actors 'choose' and 'decide'.[14]

3. The 2 × 2 prisoners' dilemma game: A structural analysis

We use our theory of social interaction to structurally characterize and analyze the prisoners' dilemma game (Rapoport, 1960; Rapoport and Chammah, 1965; David, 1970). The various $4m+2$ components of $\Gamma(\mathcal{M})$, $m=2$, must be specified empirically in order to fully represent and analyze the particular social situation.

An interaction situation with partial opposition is characterized by some degree of correlation of the preference structures R^1 and R^2 (in the case of the two actor problem), that is $0 < \gamma(R^1, R^2) < 1$. In the specific case of the prisoners' dilemma, represented below, opposition in preference structures occurs at the highest and lowest levels while concordance obtains at the middle levels.

In the two actor/two alternative/complete control (see p. 41) system $\mathcal{M} = \{1, 2\}$. The 2-actor action structure is $\Gamma(\mathcal{M}) = (\mathcal{M}, E^1, E^2, d^1, d^2)$, where $E^i = (\Theta, R^i, A^i, H^i)$, $i=1, 2$ with $A^i = \{1, 2\}$ (two alternatives), $\Theta = \{w, x, y, z\}$, $H^1(1) = \{w, x\}$, $H^1(2) = \{y, z\}$, $H^2(1) = \{w, y\}$, $H^2(2) = \{x, z\}$ and

$R^1: y > w > z > x$

$R^2: x > w > z > y$

$\sigma(R^1, R^2) =$	0	1	2	3
0	∅	∅	∅	{y}
1	∅	{w}	∅	∅
2	∅	∅	{z}	∅
3	{x}	∅	∅	∅

are the preference relations and their mutual level structure matrix (here $x>y \leftrightarrow (x, y) \in R^i$). This is a partial opposition situation in that there is some degree of correlation between the preference structures R^1 and R^2 (see Note 11).

We assume that R^1 and R^2 reflect individual or personal evaluations of the outcomes. Then, in the case of pure self-orientation (see p. 43) on the part of each actor, the rankings R^1 and R^2 would be unaffected by knowledge of one another's preferences, and each actor would be likely to use the conservative maxmin procedure ($d_1 = d_2 =$ maxmin).[15] This yields the action preferences $Q^1_{Mm} = \{(2, 1)\} = Q^2_{Mm}$ (action 2 preferred by each actor) with the resulting outcome z.[16] Note that had each chosen alternative 1, the outcome would have been w, which each prefers to z. On the other hand, should one actor choose 1 with this expectation, the other could choose his second alternative with the result that he attains his highest ranked outcome and the former actor, his lowest ranked outcome. This paradox, that actors behaving according to rational selfishness, may each obtain payoffs less than that obtainable by 'irrational' cooperative behavior has been discussed elsewhere as the prisoners' dilemma (Davis, 1970, pp. 93–103, 128–131; Rapoport, 1960, pp. 173–181). The structural nature of the paradox is clear. It arises from a particular combination of preference relations, action-outcome structure, and decision procedure.

Specifically, the paradox arises when the AOP-structures and decision procedures generating a preference order over action alternatives result in the following inconsistency: certain actions $\alpha \in A^1$ and $\alpha \in A^2$ are preferred by actors 1 and 2 and $H^1(\alpha) \cap H^2(\alpha) = \Theta(\alpha, \alpha)$, but there exist $\beta \in A^1$ and $\beta \in A^2$ with $H^1(\beta) \cap H^2(\beta) = \Theta(\beta, \beta)$ preferred to $\Theta(\alpha, \alpha)$ by each actor. In short, $(\alpha, \beta) \in Q^i$ but $(\Theta(\beta, \beta), \Theta(\alpha, \alpha)) \in R^i$ for each actor $i \in \{1, 2\}$. Note that for a maxmin decision procedure, because $\Theta(\alpha, \alpha)$ is the maxmin outcome, were one actor to choose alternative β in the expectation of attaining $\Theta(\beta, \beta)$, the other could choose alternative α which would lead to an outcome more preferred by the latter but less preferred by the former than the maxmin outcome.

If we write for outcomes $\Theta(\alpha^*, \beta^*)$, $\Theta(\alpha, \beta)$, actor $i \in \{1, 2\}$ ($\Theta(\alpha^*, \beta^*)$, $\Theta(\alpha, \beta)) \in R^i \Leftrightarrow \Theta(\alpha^*, \beta^*) \xrightarrow{i} \Theta(\alpha, \beta)$, we can represent the prisoners' dilemma structurally as follows:

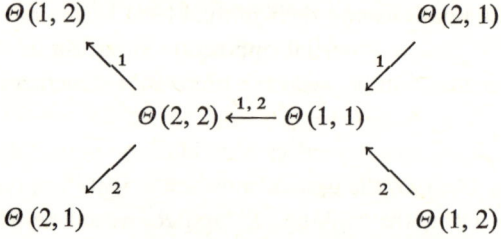

Structural characterization of PDG

In words, actor 1 most prefers $\Theta(2, 1)$ and least prefers $\Theta(1, 2)$ while the ordering is reversed for actor 2, and yet both prefer $\Theta(1, 1)$ to $\Theta(2, 2)$. Thus, although alternative 2 is a dominant one since its outcomes dominate the outcomes of the other alternative (1) for each action of the other player, there exists at least one outcome $\Theta(1, 1)$ which is better than the likely or determined outcome of the dominant alternatives for the two actors ($\Theta(2, 2)$).

The dilemma characterized above may occur also in the case the actors have altruistic orientations toward one another, and each tries to choose the alternative giving the best outcome to the other (maximizing procedure), but the AOP-structures diagrammed on page 46 obtain. Each actor chooses action 2, giving a result $\Theta(2, 2) = z$, which is worse than that they had hoped for and worse than that possible through coordination of their actions. *The dilemma* and the source of the frustration lie in the structure of the situation. In general, it is possible in many decision situations to construct or generate a prisoners' dilemma type of problem with the structural properties of the system represented above (Buckley and Burns, 1973).

4. Resolutions and Social Control of Prisoners' Dilemma

Our theory, as mentioned earlier, offers a general characterization of a social interaction system. Changes in the components of the interaction structures $\Gamma(\mathcal{M})$ or in meta-processes or constraints which affect the components of $\Gamma(\mathcal{M})$ result generally in a transformed structure $\Gamma(\mathcal{M})^*$.

Using the theory, we show below how the prisoners' dilemma situation, as an interaction system, is resolved and controlled through certain well-defined changes in (or social controls on) the components of the system. In particular, we examine the role of institutionalized constraints (exogenous controls) as well as those of established and emergent relationships between the actors (endogenous controls) in altering preference structures, action alternatives, and decision procedures in such a way as to transform the prisoners' dilemma situation. We also indicate some of the conditions under which such changes or constraints are likely to obtain.

A. *Changes in Preference Structures as a Function of Interpersonal Sentiments and Influence Processes*

In characterizing the prisoners' dilemma, we assumed that R^1 and R^2 reflected individual or personal evaluations of the outcomes. In the case the actors have pure *self-orientations*, their rankings would be unaffected by knowledge of one another's preferences. On the other hand, whenever the actors have *mutual self/other orientations* (see p. 43) (as a function, for example, of role relationships or interpersonal sentiments), knowledge of one another's preferences would lead to predictable changes in their preference structures. Both actors would be motivated by their own personal preferences as represented by R^1 and R^2 as well as by their concerns for one another. To the extent that each considers both preferences of equal importance, they are led to search for *mutually satisfying outcomes*. These are the outcomes lying highest along the main diagonal of the structural matrix $\sigma(R^1, R^2)$ (w in the example) rather than those characterized by maximum opposition (lying in the upper right or lower left-hand corners of $\sigma(R^1, R^2)$ (here x and y)). As a result of the social orientations of the actors in relation to one another (a meta-process), *their preference relations are effectively changed to*

$$\bar{R}^1: w > z > \{y, x\} \qquad \bar{R}^2: w > z > \{y, x\}$$

which, when combined with maximizing decision procedures, yields the action orders $\bar{Q}_M^1 = \{(1, 2)\} = \bar{Q}_M^2$ and the most preferred outcome w in $\bar{R}^1$ and $\bar{R}^2$.

If each actor has a *negative other orientation* (see p. 43), he would rank outcomes in terms of the harm or costs they bring to the other actor. In this instance, the preference relations are transformed (due to the orientations) to patterns $\underline{R}^1$ and $\underline{R}^2$

$$\underline{R}^1: y > z > w > x \qquad \underline{R}^2: x > z > w > y$$

Either a maximizing or a maxmin decision procedure results here in the action preference relations $\underline{Q}^1 = \{(2, 1)\} = \underline{Q}^2$. In view of the orientations of the actors and the relations $\underline{R}^1$ and $\underline{R}^2$, the outcome z does not, in this case, create the sense of paradox associated with the same outcome in the case of pure self-orientations discussed above.

In this example, we have shown how social orientations between actors operate as meta-processes to transform preferences, decision procedures, and interaction patterns. More complicated and realistic examples provide opportunities to make use of these concepts more fully as well as to illustrate other types of meta-processes.

While the discussion above has focused on the influence of established social relationships (and orientations) on preference structures, many social interactions entail dynamic processes resulting in change of sentiments or, in some instances, in activation of sentiments or social norms (e.g., through personal appeals and moral persuasion) to the effect that preference structures are changed in the manner described above.

Our theoretical approach suggests that it is important in social research to focus on the relationship between actors on *multiple levels*, in particular, to investigate influence processes, for example, with respect to preference structures: actor i attempts to change j's R^j and vice versa. Perceived status or competence (Berger and Conner, 1969; Berger et al., 1972) and activation or establishment of interpersonal sentiments, obligations, or social norms are a few of the factors that play a role in influence processes and, therefore, in changing preference structures. Thus, a social or religious movement which persuades persons touched by it to rank 'cooperation' over 'competition' and 'generosity' over 'exploitation' in effect alters preference structures such as those characteristic of the prisoners' dilemma game.

In general, changes in the perception of action outcomes or in the evaluation bases of preference can bring about the same result: a restructuring of preference relations.

B. *Changes in Action Sets and Action-Outcome Relations*

By considering additional actions (a meta-process expanding A^1 and A^2)[17] or altering action-outcome relations (H^1 and H^2), the actors may succeed in resolving the prisoners' dilemma. In the following example, the actors have introduced new options with an outcome which they both rank over all other outcomes:

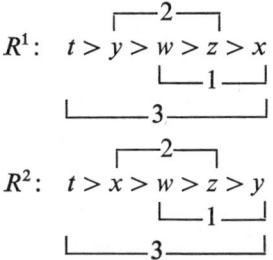

Whenever one or both of the actors find outcomes such as x and y unacceptable or, in general, are unable to realize his or their goals (which in the case discussed on pp. 49–50 include one another's goals), one or both can be expected to try to *generate new action alternatives with more satisfactory outcomes* (that is, a meta-process ensues).

Such activity results in the expansion of the sets of action alternatives (A^1 and/or A^2 expand to $\bar{A}^1$ and $\bar{A}^2$) and of the interaction matrix. This may produce an outcome satisfying the goals and expectations (evaluation criteria) of both actors, as in the transformation diagrammed above. In a word, they transform the PD interaction situation into one, for all intents and purposes, of greater or more effective concordance.

Whether the actors have the capacity to accomplish this transformation will be a function of certain institutional and interpersonal constraints (see discussion on pp. 44–46). In the absence of such constraints, actors with a positive or cooperative relationship can satisfactorily redefine their situation, establishing *non-prisoners' dilemma type interaction systems*. On the other hand, such resolutions are especially difficult in the case of a mutually hostile relationship or, in some instances, even in the case of a relationship of indifference (joint pure self-orientations). Hostile or indifferent actors would be less likely to cooperate in transforming the situation collectively than actors related more positively. Rather, mutually hostile actors would tend, in the absence of constraints, to expand the game into one with greater opposition. Clearly, the social relationships among the actors (as reflected in their orientations toward one another) play a crucial role in evaluations, decision-making, and interaction patterns. By investigating such relationships, one can anticipate emergent properties of social interactions.

Actors also expand the sets A^i by considering combinations or more complex forms of the actions already included in A^i. Statistical combinations ('mixed strategies') of elements in A^i entail such an expansion. (In the prisoners' dilemma game, statistical strategies are of no relevance since each player already has a strategy that is 'best' (dominant) regardless of what the other does).

Other examples are contingency or reciprocity actions whereby each actor i considers his actions in A^i in response to or conditional on another actor's (j's) response (in A^j). A^i is thus expanded to include conditional or contingent types of actions such as reciprocating action ('I'll do 1 if you do 1').

Pursuing this line of thought, Nigel Howard (1966, 1971) has conceptualized the construction of a 'meta-game' that goes beyond the given action alternatives or strategies in A^1 and A^2.[18] In his model, preference structures with respect to the component acts of the conditional strategies and decision procedures remain the same (these are questionable assumptions, see Burns (1973), but which in no way detract from the logic of his argument). The result is that in the expanded game, both mutual cooperation (1, 1) and mutual non-cooperation (2, 2) are shown, in game theory terms, to be equilibrium points.

C. *Use of Collective Decision Procedures*

The individual maxmax and maxmin procedures, when applied to the AOP-structures on p. 46, result in the prisoners' dilemma paradox. However, with R^1 and R^2 as well as with A^1 and A^2 constant, there exist other decision procedures, some of which resolve the dilemma. Of particular interest are collective decision procedures.[19]

By a collective decision procedure, d^c, we mean one in which actors in $\mathcal{M}$ combine or aggregate individual preferences (e.g., through voting[20]) to obtain a social preference structure over outcomes (R^c) or over multiple actor actions (Q^c), selecting the most preferred alternative ranked in this way. The selected m-action vector, α, with outcome(s) $\Theta(\alpha)$ is prescribed (binding) for all actors in $\mathcal{M}$.[21]

Collective decision procedures such as majority as well as authoritative rule enable a group of actors to combine their preferences and decisions so as to obtain a better outcome than that achievable through individual or independent decision-making. Thus, in the case of the prisoners' dilemma, if the actors link or coordinate selection of alternative 1 (for both), they obtain w, and outcome preferred to the z outcome resulting from independent optimizing, maxmax, or maxmin procedures.[22]

Russell Hardin (1971) has shown that if in an m-actor prisoners' dilemma game, the m-actors vote on outcomes, the cooperative outcome $\Theta(1, 1, ..., 1)$ (= 'w' in the 2-person case) is a Condorcet choice[23] and would prevail in an election against all other outcomes. In particular, using a collective procedure rather than m independent decision procedures, the actors would, in pairwise voting, choose outcome $\Theta(1, 1, ..., 1)$ over all other outcomes. That is, the Pareto optimum outcome would be obtained over other outcomes such as the 'equilibrium outcome' of game theory, $\Theta(2, 2, ..., 2)$ (= 'z' in the 2-person case).

Using individual decision procedures, i may have little or no influence, other than through his own choice α_i over the multi-action vector α and the outcome $\Theta(\alpha)$. By means of a collective procedure, actor i and the other members of $\mathcal{M}$ gain greater control over α and $\Theta(\alpha)$.[24] In particular, in the case of the two person prisoners' dilemma game, the actors can achieve the Pareto optimum $\Theta(1, 1)$ through collective choice of $\alpha = (1, 1)$.

Clearly, collective decision procedures are rational in many social settings, especially those with the properties of the prisoners' dilemma (Buckley and Burns, 1973), since they enable actors to achieve a better outcome than would otherwise be possible.

'Selection' of collective decision procedures rather than individual procedures entails a meta-process. Moreover, it indicates a particular opportunity structure (action possibilities to utilize collective procedures) as well as social controls or interpersonal sentiments (self/other or other orientations) that make the collective choice effective, that is, prescriptive or binding.[25]

Obviously, one actor i cannot use a collective procedure d^c while j uses an individual procedure d^j. Hence, whether based on normative prescription, long-term calculation, or mutual concern about one another, concordance about d^c among the actors is necessary for an effective collective decision process. Put in other words, the actors must agree *at the meta-process level* about the use of a collective procedure in a social setting and, in particular, in regard to resolving dilemmas or conflicts.

Concordance about the use of collective procedures in a group usually depend on certain social conditions such as interpersonal sentiments, cultural understandings, or social norms prescribing such procedures and enforcing their results. Social influence and constraint operate at the meta-level selecting and making effective a collective decision procedure. Without such social controls, either actors would make decisions individually or, while agreeing collectively on an alternative, some or all of the participants could cheat one another by not implementing the vote or negotiated agreement. In a word, there is a *higher order or meta-prisoners' dilemma game*, the resolution of which depends on the social context: the nature of interpersonal relationships, normative controls, etc.

D. *Normative and Interpersonal Controls*

Social controls in interaction systems operate on the m-actor action structure $\Gamma(\mathcal{M})$, in particular, on preference structures, action sets (opportunity sets) and decision procedures.

(i) As already suggested in our discussion in (A), interpersonal relationships and social norms affect preference structures, encouraging or prescribing actors to take one another into account. For example, actors in role positions (such as doctors and nurses, teachers, public servants, parents, etc.) in relation to other persons, classes of persons, or roles (patients, students, citizens, children, etc.) are normatively obligated to pursue their interests. That is, their behavior is expected to be guided in particular social settings (a 'service context') by the evaluation and decision base defined earlier as 'positive other-orientation'. Incumbents of such roles are usually socialized to have the proper other-orientation and are also subject to social controls maintaining or reinforcing the orientations (Burns and Cooper, 1971, pp. 319–326).

(ii) Interpersonal and institutional constraints determine actors' access to resources and their action possibilities and therefore, define the limits within which they make choices: the viable action sets A^i in a given context. Thus, normative controls may restrict the range of permissible conduct, prohibiting actions through which actors gain advantages at the expense of others or of a collective interest: for example, prescribing 'cooperative acts' or proscribing exploitative patterns (2, 1) or (1, 2) in prisoners' dilemma situations. This may be achieved through group constraints making such acts impossible or through group sanctions restructuring payoffs (see (A) above).

(iii) As discussed in (C), opportunities to use collective decision procedures depend on interpersonal and institutional conditions. In game theory, this question is approached in terms of the conceptual distinction between 'cooperative games' and 'non-cooperative games'. In cooperative games, the actors can communicate and make binding (enforceable) agreements. In non-cooperative games, they cannot (see Notes 26 and 27). In a word, the 'rules of the game' (constraints) under which actors interact are conceptually distinguished (see Section 2E).

In our sociological theory, there is an explicit focus on the context and, in particular, on the social processes that constrain or structure interaction and social relationships (Burns, 1973; Burns and Cooper, 1971). We predict that, in general, there will be a close connection between *convergent preference structures, use of collective decision procedures and social conditions allowing for or facilitating 'cooperative games' (with communication and binding agreements)*.

(1) The social controls which contribute to such correlations may be based on the direct relationships among the actors ('endogenous processes of control' (Burns, 1973)). Actors with enduring and multiplex ('many-stranded') social relationships are likely to take one another into account (self/other or other-orientations), to communicate with one another, and to cooperate in achieving jointly satisfying outcomes.[26] They apply sanctions (and constraints) to one another. That is, the form and content of the interactions are established and maintained by the actors themselves – and hence are distinguishable from more institutionalized forms (discussed below) by self-regulating mechanisms. Each actor involved has a personal interest in sustaining the relationships (see Note 26) and, therefore of avoiding actions such as cheating or exploitation which would permanently disrupt the relationship.

(2) Social controls may depend on institutional and ideological conditions exogenous to the relationship (Burns, 1973). The weaker the sentiments among the actors, the more controls must be accomplished through such external mechanisms, making, for example, voting obligatory and its results binding. In general, the actors are socially constrained to transact with one another in a particular manner (such as to have self/other orientations or to utilize collective decision procedures).[27] Thus, although the concordant (publically expressed) preference structures or collective procedures may appear to be instigated by the actors themselves,

they often derive from compliance with externally maintained social norms or role obligations.

As the brief discussion above suggests, the social context of interaction is at least as important as the material (the context or specific 'gains' and 'losses' of transactions) and scarcely can be represented by a scalar-valued utility function. It is here that a multi-dimensional theory of social behavior may serve as a valuable tool for analysis and empirical-research.

APPENDIX

The Comparison of Digraphs: The Mutual Level Structure Matrix and The Concordance Index (see Burns and Meeker, 1973)

Let D be a diagraph with points $V = \{v_1, v_2, ..., v_N\}$ (an irreflexive relation on V, see Harary *et al.* (1966)). A sub(di)graph, S, of D, having points $V' \subset V$ is said to be a *strong component* of D if 1) every pair of points of V' are mutually reachable, and 2) S is maximal with respect to this property. The strong components of D partition V. The *condensation digraph*, D^*, of D is the digraph having the set $\mathscr{S} = \{S_1, S_2, ..., S_K\}$ of strong components of D as points with $(S_i, S_j) \in D^*$ if, and only if, there exist $v_i \in S_i$, $v_j \in S_j$ with $(v_i, v_j) \in D$. D^* is an acyclic graph.

Let $\mathscr{S}_0$ denote the subset of strong components of D having indegree 0 in D^* (the strong components, S, having the property that $(S', S) \notin D^*$ for any $S' \in \mathscr{S}$) and define $l: \mathscr{S} \to \{0, 1, 2, ..., N-1\}$ by setting $l(S)$ equal to zero for those S in $\mathscr{S}_0$ and equal to the length of the longest path from $\mathscr{S}_0$ to S when S is not in $\mathscr{S}_0$. l is a level assignment function for D^* with the property that if $(S_i, S_j) \subset D^*$, $l(S_i) < l(S_j)$.

We extend l to a level assignment, l_D, for D by defining $l_D(v) = l(S)$ when S is the unique strong component of D containing v.

In order to compare two digraphs D and D' with the same set of points V, we define the sets

(A, 1) $\quad L(D, i) = \{v : l_D(v) = i\}$
$\qquad L(D', i) = \{v : l_{D'}(v) = i\}$, $\quad i = 0, 1, 2, ..., N-1$

and the $N \times N$ set-valued matrix $\sigma(D, D')$ whose entries are the sets $\sigma_{i,j}(D, D') = L(D, i) \cap L(D', j)$.

The matrix $\sigma(D, D')$, called the *mutual level structure matrix* of the digraphs D and D', has the following properties which follow directly from the definition.

(A, 2) $\quad \bigcup_{j=0}^{N-1} \sigma_{i,j}(D, D') = L(D, i), \quad i = 0, 1, \ldots, N-1.$

(A, 3) $\quad \bigcup_{i=0}^{N-1} \sigma_{i,j}(D, D') = L(D', j), \quad j = 0, 1, \ldots, N-1.$

(A, 4) $\quad$ if $m = \max_v l_D(v)$ and $m' = \max_v l_{D'}(v)$, then
$$\bigcup_{K=0}^{N-1} \sigma_{i,K}(D, D') = \bigcup_{K=0}^{N-1} \sigma_{K,j}(D, D') = \phi \text{ for } i > m \text{ and } j > m'.$$

(A, 5) $\quad$ if D and D' assign the same levels to each $v \in V$, $\sigma(D, D')$ is a diagonal (set valued) matrix.

In order to construct a scalar indicator of agreement (or disagreement) between D and D', we take the matrix $\sigma^*(D, D')$ obtained from $\sigma(D, D')$ by replacing each element $\sigma_{ij}(D, D')$ (a set) by its cardinality (the number of distinct points in the set) denoted by $\#\sigma_{i,j}(D, D')$.

Definition. Let D and D' be two digraphs over the same set of N points. Then the *concordance index* of D and D' is

$$\gamma(D, D) = 1 - \frac{3}{N^3 - N} T(\sigma^*(D, D'))$$

where $\sigma^*(D, D')$ is the $N \times N$ matrix whose entries are

$$\sigma^*_{i,j} = \#(L(D, i) \cap L(D', j)), \quad 0 \leqslant i, j \leqslant N - 1.$$

and

$$T(\sigma^*) \equiv \sum_{i,j=0}^{N-1} (i - j)^2 \sigma^*_{i,j} \leqslant \frac{N^3 - N}{3}.$$

The concordance index of D and D' satisfies the following:

(1) $\quad 0 \leqslant \gamma(D, D') \leqslant 1$

(2) $\quad \gamma(D, D') = 1$ if, and only if, D and D' assign the same levels to each $v \in V$.

(3) $\quad \gamma(D, D') = 0$ if, and only if, D and D' are each acyclic, unilateral and directional duals.

Note that the concordance index, as defined here, is a generalization to digraphs of the familiar Spearman's rank correlation used in the comparison of rank orders.

University of New Hampshire

NOTES

* We are grateful to Hayward Alker, Jr., Walter Buckley, and Russell Hardin for their helpful comments on an earlier draft of this paper.

[1] The more technical formulation of this theory appears in Burns and Meeker (1973, 1972).

[2] The actors may be individuals, groups, parties, governments, or in general, any organized group characterized by possession of goals or objectives in the interaction situation and by the capacity to carry out collective acts. There are, of course, 'internal social processes' in the case i and j are groups.

[3] This is somewhat restrictive and could be replaced without undue complications by the cartesian product of sets having a partial order.

The various coordinates describing outcomes are not assumed to be of equal importance to an actor (see Burns and Meeker, 1973).

[4] Elsewhere (Burns and Meeker, 1972), we consider actors' differential perceptions of the situation, in particular, of action outcomes.

[5] For example, consider the set $\Theta = (w, x, y, z)$ symbolizing multi-dimensional outcomes with, in many instances, components of a non-quantitative character. There may be a partial or even a weaker ordering over the entities. The following are several examples of preference structures (note that the 'x' and 'y' in (ii), (iii), (iv), and (v) are non-comparable, as are 'z' and 'y' in (iii) and (iv)).

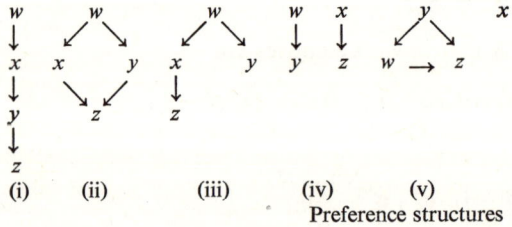

Preference structures

[6] Elsewhere, we analyze the problem of multi-dimensional evaluation as summarized by R^i (Burns and Meeker, 1973).

[7] x preferred to $y \leftrightarrow (x, y) \in R^i$.

[8] Since we can consider R as a digraph with level assignment l_R, the condition above can be weakened by replacing '$(u, v) \in R$' by '$l_R(u) \leqslant l_R(v)$'. This relation we denote by Q_{OP}.

[9] Social constraints and norms may operate as selection devices (see Sub-section E below and 4D).

[10] Yet other examples of meta-processes relevant to our model are selection of feasible action alternatives for consideration and selection of a preference structure over outcomes or the system generating the preference structure.

[11] In this discussion the actors are assumed to know one another's orientations and preference structures. This knowledge may be based on shared cultural understandings, role or previous social relationships, or even information given each of them by an authority (e.g., the experimenter in small group studies).

[12] $R_1{}^j = \Theta \times \Theta$ and $\sigma(R_1{}^1, R_1{}^j)$ becomes a column vector expressing only 1's evaluations.

[13] B. Lieberman (in Buchler and Nutini (1966, p. 100)) points out: Incest taboos may be interpreted as quite successful attempts to prevent family members from expressing their sexual preferences for each other. The assignment of certain economic activities to certain families or subgroups of a particular society serves to prevent other members of the society from expressing their preferences for desirable occupations, with the resulting economic conflict.... The traditions and customs of both Western and non-Western societies prevent conflicts from arising that stem from contradictory preference patterns, and are, therefore, conflict resolution mechanisms. These conflicts may exist in many domains of behavior: the economic domain, the family-personal domain, the domain of permissable sexual behavior, and the domain of political behavior.

[14] Buchler and Nutini (1968, p. 8) distinguish between the anthropologist's general interest in ground rules, which structure the cultural framework within which decision-making occurs, and the game theorist's interest in strategy rules or rules for playing games intelligently, which guide choices among the options the cultural framework allows:

Just as the anthropologist often ignores strategy rules and the context in which various moves may be employed, the mathematical game theorist and the experimentalist often ignore in the name of experimental control, the rules that structure the framework within which decision-making occurs.

Making somewhat the same conceptual distinction, Howard (1963, p. 436) has argued:

In short, I am suggesting that socio-cultural principles may be thought of as setting the limits of acceptability, but that the degree of freedom permitted an individual (i.e., the number of alternative choices possible within those limits) differs from society to society, and from activity system to activity system within each society. To the degree that choices are limited by socio-cultural principles, one would expect socio-cultural explanation to suffice; conversely, to the degree that they permit choice, one would expect psychological principles to be required in addition.

[15] The same result obtains for individual maximizing or sure-thing procedures.

[16] However, even in one-shot games where the actors do not know one another and, in theory, should *only* be concerned with their individual gains and losses, they, as socialized humans, bring into the situation internalized norms such as 'consideration for others' and certain cooperative tendencies that insure that not all of them would play in the selfish manner described here (Shubik, 1970) (see p. 48ff).

[17] This is provided that there are no constraints preventing them from doing so (see pp. 44–46).

[18] Given the two basic alternatives {1, 2}, actor 1 may conceptualize and consider the following conditional strategies:

1*　　　choose 1 regardless of what actor 2 chooses or is expected to choose.
2*　　　choose 2 regardless of what actor 2 chooses or is expected to choose.

3   choose 1 or 2 according to what the other does or is expected to do.
4   choose 1 or 2 in opposition to what the other does or is expected to do.

The result is that in place of A^1 we find $\bar{A}^1$ consisting of 1*, 2*, 3, and 4. Similarly actor 2 may transform A^2. Now, 2 will have 16 conditional strategies related to actor 1's possible choices. To each of 1's assumed four choices, 2 can respond with either 1 or 2 in four different ways. The result is a 16 × 4 'meta-game' with 64 outcomes.

[19] Individual satisficing procedures (Burns and Meeker, 1973; March and Simon, 1958) where the goals or expectations are set for the Pareto optimum (w) would also resolve the dilemma. Such expectations differentiate w from z and y for actor 1, and w from z and x for 2. The actors might reason that, given the nature of the interaction situation, the best strategy would be to act so as to be satisfied with the Pareto optimum w, although 1 prefers x to w and 2 prefers y to w. Given their expectations that outcomes x or y are unstable and unlikely in any repeated process and that they are both reasonable persons, each would rank 1 over 2, giving the (1, 1) interaction pattern.

[20] Voting procedures range from plurality, majority, two-thirds rule, to the most stringent unanimous rule.

[21] The process of considering and utilizing conditional acts, as discussed in (B) above, does not entail a collective procedure for obtaining Q^c, nor is a jointly preferred multi-action vector, $\alpha^* = (\alpha^*_1, \alpha^*_2)$ a binding one on the actors. They still make individual choices, i, for instance, changing α^{*i} regardless of what the other chooses or does.

[22] Re-evaluation produces new preference structures (changes in R^1 and R^2). However, the original evaluation base may persist. The result is often ambivalence and the probability that the original structure will be reactivated. On the other hand, changes in decision procedure such as adoption of collective decision procedures is not necessarily associated with a change in the underlying ranking of outcomes by individuals.

[23] In an m-actor group $\mathcal{M}$ where s and t are outcomes from the set of realizable outcomes in the game matrix, let m_{st} be the number of those in $\mathcal{M}$ who prefer s to t, m'_{st} be the number who are indifferent to whether outcome s or t obtains, and m_{ts} be the number of those preferring t to s (Hardin, 1971). Clearly,

$$m_{st} + m_{ts} + m'_{st} = m$$

DEFINITION (Hardin, 1971): s is a *strong* Condorcet choice if it is preferred by a majority in $\mathcal{M}$ to every $t (\neq s)$ in Θ. In short,

$$m_{st} > m/2 \quad \text{for all} \quad s \neq t.$$

DEFINITION (Hardin, 1971): s is a *weak* Condorcet choice if it is not a strong Condorcet choice but if, for each $s \neq t$, more of those in $\mathcal{M}$ prefer s to t than t to s. This condition is simply

$$m_{st} > m_{ts} \quad \text{for all} \quad s \neq t.$$

[24] Voting procedures are only one type of collective decision procedure. Authoritative rule – decision-making by a leader, elected or not – is another. The problem of collective action (and, in particular, the prisoners' dilemma) can be dealt with by a leader who ranks (possibly taking cognizance of the 'sense of group opinion') $\Theta(1, 1, 1, ..., 1)$ highest and makes sure $\alpha = (1, 1, 1, ..., 1)$ is carried out.

[25] There is first the question of whether or not the actors have the opportunity to use collective decision procedures or, in general, to establish a 'cooperative game', that is,

a choice between $C\Gamma$ and $\bar{C}\Gamma$. Either there may be no choice, one or the other being prescribed or excluded. Or given a choice, the willingness of the actors to collectively choose $C\Gamma$ depends on the nature of the relationships among them or the social controls operating on them (see pp. 54–56).

[26] The perceived durability of a positive relationship provides a *temporal context* to any specific interaction situation. Betrayal, cheating, or exploitation of another will have more than immediate or circumscribed consequences (specific 'gains' or 'losses'). This condition serves to constrain such behavior. Similarly, a network of relationships in an enduring group (stable networks) make up a *structural context* to any given interaction situation. Chicanery or exploitation would have more than immediate or fully perceived consequences (see Burns, 1973).

[27] In terms of the metaphor of the prisoners' dilemma game, the 'district attorney' as an external power or authority structures to some degree the prisoners' relationship or its context (the 'rules and conditions of interaction'). (Impersonal institutional arrangements also structure relationships (see Section 2E). (a) An external power or authority, A, may structure the relationship between the actors so as to encourage them, as in the PDG, to be distrustful and uncooperative (e.g., competitive or antagonistic toward one another), in short, to create or maintain an 'Hobbesian situation'. This is the objective of the classic 'divide and rule principle'. A does this in two ways: (1) He discourages or prevents communication, collective decision-making, and binding agreements between the actors. And (2) he determines or influences the structure of payoffs in the direction of the social preference structure on p. 48, thereby increasing the likelihood of non-cooperative behavior between the actors (especially under the conditions of (1)). (b) On the other hand, the external power or authority may structure the relationship so as to encourage mutual trust and cooperativeness by (1) facilitating communication, collective decision-making, and binding agreements and (2) by determining or influencing the structure of payoffs in a direction away from the prisoners' dilemma preference structure – e.g. toward a structure with $\Theta(1, 1)$ as the most preferred outcome for each actor (see Buckley *et al.*, 1973). Shubik (1970) points out that in much of everyday political, economical and social life, prisoners' dilemma situations are dealt with by third parties (the state or social agents) acting as mediators and enforcing or policing agreements.

BIBLIOGRAPHY

Arrow, K.: 1963, *Social Change and Individual Values*, 2nd ed., Wiley, New York.

Berger, J. et al.: 1972, 'Status Characteristics and Social Interaction', *American Sociological Review* 37 241–255.

Berger, J. and Connor, T.: 1969, 'Performance Expectations and Behavior in Small Groups', *Acta Sociologica* 12 186–198.

Buchler, I. R. and Nutini, H. G. (eds.): 1968, *Game Theory in the Behavioral Sciences*, University of Pittsburgh Press, Pittsburgh.

Buckley, W.: 1967, *Sociology and Modern Systems Theory*, Prentice-Hall, Englewood Cliffs, New Jersey.

Buckley, W., Burns, T., and Meeker, L. D.: 1973, 'A Structural Theory of Collective Action', unpublished manuscript.

Buckley, W. and Burns, T.: 1973, 'Situational Rationality', in preparation.

Burns, T.: 1973, 'A Structural Theory of Social Exchange', *Acta Sociologica* 16, 188–208.

Burns, T. and Cooper, M.: 1971, *Value, Social Power, and Economic Exchange*. Samhallsvetareforlaget, Stockholm.

Burns, T. and Meeker, L. D.: 1973, 'A Mathematical Model of Multi-dimensional Evaluation, Decision-making, and Social Interaction', in J. Cochrane and M. Zeleny (eds.), *Multiple Criteria Decision-Making*, University of South Carolina Press, Columbia, S.C., 1973.

Burns, T. and Meeker, L. D.: 1972, 'A Mathematical Theory of Social Interaction', unpublished manuscript.

Burns, T.: 1972, 'Cooperation and Conflict', paper presented at the Symposium on *New Directions in Theoretical Anthropology*, SUNY, Oswego, New York.

Davis, M. D.: 1970, *Game Theory*, Basic Books, New York.

Emshoff, J. R.: 1970, 'Model of Prisoner's Dilemma', *Behavioral Science* **15** 304–317.

Harary, F., Norman, R., and Cartwright, D.: 1965, *Structural Models: An Introduction to the Theory of Directed Graphs*, Wiley, New York.

Hardin, R.: 1971, 'Collective Action as an Agreeable n-Prisoners' Dilemma', *Behavioral Science* **16** 472–481.

Howard, A.: 1963, 'Land, Activity Systems, and Decision-Making Models in Rotuma', *Ethnology* **2** 407–440.

Howard, N.: 1966, 'The Theory of Metagames', *General Systems* **11** 167–186.

Howard, N.: 1971, *Paradoxes of Rationality: Theory of Metagames and Political Behavior*, MIT Press, Cambridge, Mass.

Rapoport, A.: 1960, *Rights, Games and Debates*, Univ. of Michigan, Ann Arbor.

Rapoport, A. and Chammah, A. M.: 1965, *Prisoner's Dilemma: A Study of Conflict and Cooperation*, University of Michigan Press, Ann Arbor.

Shick, F.: 1972, 'Democracy and Interdependent Preferences', *Theory and Decision* **3** 55–76.

Shubik, M.: 1970, 'Game Theory, Behavior, and the Paradox of the Prisoner's Dilemma: Three Solutions', *Journal of Conflict Resolution* **14** 81–193.

D. MARC KILGOUR

ON 2×2 GAMES AND BRAITHWAITE'S ARBITRATION SCHEME

The problem of arbitration of two-person non-zero-sum games has been given much attention by game theorists. Even when meaningful utility comparisons between the two players are possible, a difficulty has arisen. For, if utilities are measured on an interval scale, an arbitrated result should not be affected by the application of different positive linear transformations to the payoffs of the two players. A natural solution is to distinguish certain transformations which would 'normalize' the payoff matrices, and thus permit direct numerical comparison of utilities of different players. Then arbitration is reduced to a problem of cooperative bargaining, which can be solved by the use of lines of constant relative advantage. Braithwaite [1] has proposed such a method. As described by Luce and Raiffa ([2], pp. 148–150), Braithwaite's normalizing transformation operates by equating the gains of the two players when each performs a certain change of strategy. In essence, Braithwaite's arbitration scheme rests on a particular 'symmetrization' of the game.

We shall consider the application of Braithwaite's method to 2×2 non-zero-sum games, the simplest games for which an arbitration problem exists. Firstly, conditions on the payoff matrices will be formulated which are necessary and sufficient for the applicability of Braithwaite's arbitration scheme. Secondly, the effect of arbitrary monotone (increasing) transformations on the applicability or inapplicability of Braithwaite's scheme will be studied. Finally, each of the games in Rapoport and Guyer's taxonomy of 2×2 games [3] will be classified according to the sense, if any, in which it can be arbitrated by Braithwaite's method.

The game Γ, with players 1 and 2, has the following representation:

Γ:

1 \ 2	$y=1$	$y=0$
$x=1$	a_1 a_2	c_1 b_2
$x=0$	b_1 c_2	d_1 d_2

For $i=1$ and 2, define

$$\alpha_i = d_i - b_i$$
$$\beta_i = d_i - c_i$$
$$S_i = a_i - b_i - c_i + d_i.$$

These quantities permit the representations

(1)
$$V_1(x, y) = d_1 - \beta_1 x - \alpha_1 y + S_1 xy$$
$$V_2(x, y) = d_2 - \alpha_2 x - \beta_2 y + S_2 xy$$

where $V_i(x, y)$ denotes player i's expectation in Γ given that the (mixed) strategies x and y are chosen by 1 and 2, respectively. Let the co-player of i be $j=j(i)$. The zero-sum game Γ_i is defined by the following:

Γ_i:

$i \backslash j$	1	0
1	a_i	c_i
0	b_i	d_i

Let $\underline{z}_i$ (respectively, $\bar{z}_i$) be i's maximin (j's minimax) mixed strategy in Γ_i. We shall always assume that $\underline{z}_i$ and $\bar{z}_i$ are unique (and, in particular, undominated). Define $\underline{x}=\underline{z}_1$, $\bar{x}=\bar{z}_2$, $\underline{y}=\underline{z}_2$, and $\bar{y}=\bar{z}_1$. We call $\underline{x}(\underline{y})$ the maximin strategy of 1 (2) in Γ, and $\bar{x}(\bar{y})$ the minimax strategy of 1 (2) in Γ. We say that Γ has a saddle point for i iff Γ_i has a saddle point. Γ is of Type I iff both players have unique saddle points; Γ is of Type II iff exactly one player has a saddle point, and it is unique; and Γ is of Type III iff neither player has a saddle point. If Γ has no equalities within the payoff matrix of each player, then Γ is either of Type I, Type II, or Type III.

The function sgn is defined on $\mathbb{R}$ by

$$\text{sgn}(x) = \begin{cases} 1 & \text{if } x > 0 \\ 0 & \text{if } x = 0 \\ -1 & \text{if } x < 0. \end{cases}$$

The following lemma is well-known:

LEMMA: Γ_i has no saddle point iff either $\min(a_i, d_i) > \max(b_i, c_i)$ or $\min(b_i, c_i) > \max(a_i, d_i)$.

It follows from the Lemma that if Γ has no saddle point for i, then

$$\operatorname{sgn}(S_i) = \operatorname{sgn}(\alpha_i) = \operatorname{sgn}(\beta_i) \neq 0.$$

We now define a monotone transformation, f, mapping Γ to $f(\Gamma)$. Let f_1 and f_2 be any two monotone increasing functions on $\mathbb{R}$. Then the game $f(\Gamma) = (f_1, f_2)(\Gamma)$ is defined by

$f(\Gamma)$:

	1	0
1	$f_2(a_2)$ / $f_1(a_1)$	$f_2(b_2)$ / $f_1(c_1)$
0	$f_2(c_2)$ / $f_1(b_1)$	$f_2(d_2)$ / $f_1(d_1)$

We shall frequently write a_1' for $f_1(a_1)$, a_2' for $f_2(a_2)$, etc. Also, functions of the payoff matrices of $f(\Gamma)$ will be denoted $\underline{x}'$, S_i', etc.

In the game Γ, define

(2) $$\begin{aligned} B_1 &= V_1(\bar{x}, \underline{y}) - V_1(\underline{x}, \underline{y}) \\ B_2 &= V_2(\underline{x}, \bar{y}) - V_2(\underline{x}, \underline{y}). \end{aligned}$$

B_1 and B_2 are well-defined provided that Γ is of Type I, Type II, or Type III. Braithwaite's arbitration scheme depends on the application of a positive linear transformation to one player's payoff which makes $B_1 = B_2$. Since such a transformation does not affect the values of $\operatorname{sgn}(B_1)$ and $\operatorname{sgn}(B_2)$, Braithwaite's scheme can be applied only if $\operatorname{sgn}(B_1) = \operatorname{sgn}(B_2)$. (It is easy to check that this condition is not only necessary, but sufficient.) Furthermore, the multiplicative parameter of the transformation is uniquely determined iff $\operatorname{sgn}(B_1) = \operatorname{sgn}(B_2) \neq 0$. (The particular choice of additive parameter has no effect on the result.) This motivates the following definitions:

Γ is a Braithwaite (B) game iff $\operatorname{sgn}(B_1) = \operatorname{sgn}(B_2) = 1$;

Γ is an anti-Braithwaite (anti-B) game iff $\operatorname{sgn}(B_1) = \operatorname{sgn}(B_2) = -1$;

Γ is an improperly-Braithwaite (imp-B) game iff $\operatorname{sgn}(B_1) = \operatorname{sgn}(B_2) = 0$;

Γ is a non-Braithwaite (non-B) game iff $\text{sgn}(B_1) \neq \text{sgn}(B_2)$;
Γ is a generalized-Braithwaite (gen-B) game iff for every monotone transformation f, $f(\Gamma)$ is *not* a non-B game and for some monotone transformation f^*, $f^*(\Gamma)$ is either a B game or an anti-B game.

Braithwaite has attempted to justify his choice of normalizing transformation only for B games. For this reason, and for convenience, we distinguish between B and anti-B games – even though the arbitration scheme is equally well-suited to both. Obviously, every Type I, II, or III game belongs to exactly one of the classes B, anti-B, imp-B, and non-B, which we call primary classes. The concept of gen-B games is a device which facilitates the study of the effect of monotone transformations on Braithwaite's arbitration scheme. Obviously, a gen-B game cannot be non-B. We shall discover which of the games of the other three primary classes are gen-B. Also, we shall study the relationship between the primary class of a gen-B game and the primary classes to which its images, under monotone transformations, belong.

Now we present an example which shows that the image of a B game, under a monotone transformation, can be a non-B game. Our example is related to Braithwaite's 'Two Musicians' game (described in [2], pp. 145–146). First consider

Luke \ Matthew	$y=1$ (play)	$y=0$ (do not play)
$x=1$ (play)	2 \ 1	3 \ 7
$x=0$ (do not play)	10 \ 2	1 \ 4

The only departure from Braithwaite's example is the interchange of Luke's payoffs in (0, 1) and (0, 0). (Now Luke, the classical pianist, prefers silence to Matthew's jazz.) It is easy to check that $\underline{x}=0$, $\bar{x}=\frac{9}{10}$, $\underline{y}=\frac{1}{5}$, and $\bar{y}=1$. Now (2) yields

$$B_1 = \tfrac{99}{50}, \quad B_2 = \tfrac{36}{5},$$

so that this example is certainly a B game. But consider

	Matthew $y=1$	Matthew $y=3$
Luke $x=1$	2 / 1	3 / 9
Luke $x=0$	10 / 6	1 / 8

Note that Matthew's payoffs are unaltered, as is Luke's preference ordering. (Now Luke finds simultaneous playing so unpleasant that he strongly prefers any alternative to it, and is relatively indifferent among those alternatives.) Again, $\underline{x}=0$, $\bar{x}=\frac{9}{10}$, $\underline{y}=\frac{1}{5}$, and $\bar{y}=1$. However

$$B_1 = -\frac{9}{50}, \qquad B_2 = \frac{36}{5}$$

and our example has become a non-B game through the action of a monotone transformation.

Now we consider games of Type I. The following theorem gives the primary class of all such games:

THEOREM 1: Let Γ be a game of Type I.
(a)   If $\underline{x} = \bar{x}$ and $\underline{y} = \bar{y}$, then Γ is an imp-B game.
(b)   If $\underline{x} = \bar{x}$ and $\underline{y} \neq \bar{y}$, or if $\underline{x} \neq \bar{x}$ and $\underline{y} = \bar{y}$, let the players and strategies of Γ be re-enumerated (if necessary) so that $\underline{x} = \underline{y} = \bar{y} = 1$, and $\bar{x} = 0$. Then
  (i) Γ is an imp-B game if $d_1 < b_1 = a_1 < c_1$;
  (ii) Γ is a non-B game otherwise.
(c)   If $\underline{x} \neq \bar{x}$ and $\underline{y} \neq \bar{y}$, let the strategies of Γ be re-enumerated so that $\underline{x} = \bar{y} = 1$. Then
  (i) Γ is a B game if $b_1 < a_1 \leqslant c_1 < d_1$ and $c_2 < d_2 \leqslant b_2 < a_2$;
  (ii) Γ is an anti-B game if $(d_1 < b_1 = a_1 < c_1$ or $b_1 < a_1 < c_1 > d_1)$ and $(a_2 < c_2 = d_2 < b_2$ or $c_2 < d_2 < b_2 > a_2)$;

(iii) Γ is an imp-B game if $b_1 < a_1 < c_1 = d_1$ and $c_2 < d_2 < b_2 = a_2$;

(iv) Γ is a non-B game otherwise.

Proof: From (2), (a) follows easily. Assume that (b) holds and that the players and strategies have been re-enumerated as stated. By (2), $B_1 = b_1 - a_1$ and $B_2 = 0$, so that (i) is immediate. For (ii), observe that since (1, 1) is 1's saddle point, $b_1 \leqslant a_1 \leqslant c_1$. If $b_1 = a_1$, so that Γ is imp-B, then (0, 1) is also a saddle point for 1 if $d_1 \geqslant b_1$. Since 1's saddle point is unique, $d_1 < b_1$. But now $d_1 < c_1$, so that (1, 0) is a saddle point for 1 unless $a_1 < c_1$. This completes the proof of (b).

Finally, let (c) hold with $\underline{x} = \bar{y} = 1$. Then $B_1 = d_1 - c_1$ and $B_2 = a_2 - b_2$. Now (i), (ii), and (iii) follow easily. Suppose that $c_1 < d_1$. Then (1, 0) is a saddle point for 1 unless $b_1 < a_1$. Similarly, $b_2 < a_2$ implies $c_2 < d_2$. This proves the converse of (i). If $c_1 > d_1$, then (1, 0) is a saddle point for 1 unless $a_1 < c_1$. If also $b_1 = a_1$, then (0, 1) is a saddle point for 1 unless $d_1 < b_1$. Otherwise $b_1 < a_1$, since (1, 1) is 1's saddle point. Therefore $c_1 > d_1$ implies either $d_1 < b_1 = a_1 < c_1$ or $b_1 < a_1 < c_1 > d_1$. Similarly $b_2 > a_2$ implies either $a_2 < c_2 = d_2 < b_2$ or $c_1 < d_2 < b_2 > a_2$. Hence the converse of (ii). The converse of (iii) is proven analogously, yielding (iv).

Q.E.D.

In the case when there are no equalities within the payoff matrix of each player, the statement of Theorem 1 can be considerably condensed. If the two saddle points coincide, Γ is imp-B. If the two saddle points are in the same row or the same column, but do not coincide, Γ is non-B. If the saddle points are in opposite corners, Γ is B if neither maximin strategy dominates; Γ is anti-B if both maximin strategies dominate; and Γ is non-B otherwise. The following corollary completes our investigation of Type I games:

COROLLARY: The primary class of a Type I game is not altered by any monotone transformation.

Proof: A monotone transformation does not affect the location of saddle points nor the ordinal relations within the payoff matrix of each player. Since the classification in Theorem 1 depends only on these features, the corollary follows.

Q.E.D.

Therefore, a Type I game is gen-B iff it is either B or anti-B. Observe that Type I games possess a strong 'Braithwaite-invariance' property with respect to monotone transformations.

We now consider games of Type II. The next theorem gives the primary class of these games.

THEOREM 2: Let Γ be a game of Type II. Re-enumerate the players and strategies of Γ, if necessary, so that player I has a saddle point and $\underline{x} = \bar{y} = 1$. Define

$$U = (d_1 - c_1) - (a_1 - b_1)\left(\frac{d_2 - b_2}{a_2 - c_2}\right).$$

Then Γ is a B game iff $\text{sgn}(S_2) = \text{sgn}(U) = 1$; Γ is an anti-B game iff $\text{sgn}(S_2) = \text{sgn}(U) = -1$; and Γ is a non-B game otherwise.

Proof: It has already been noted that $S_2 \neq 0$. From (1), it follows that $\underline{y} = \alpha_2/S_2$ and $\bar{x} = \beta_2/S_2$, and that, for any x, $V_2(x, \underline{y}) = d_2 - \alpha_2 \beta_2/S_2$. Consequently, $B_2 = a_2 - d_2 + \alpha_2 \beta_2/S_2 = (a_2 - b_2)(a_2 - c_2)/S_2$. By the Lemma, the numerator of the latter expression is positive; therefore,

(3) $\quad \text{sgn}(B_2) = \text{sgn}(S_2).$

From (3) it follows immediately that Γ is not imp-B, for $\text{sgn}(B_2) = 0$ is impossible. Now, from (1) and (2),

$$B_1 = d_1 - \frac{\alpha_1 \alpha_2 + \beta_1 \beta_2}{S_2} + \frac{\alpha_2 \beta_2 S_1}{(S_2)^2} - a_1\left(\frac{\alpha_2}{S_2}\right) - c_1\left(\frac{S_2 - \alpha_2}{S_2}\right)$$

$$= \frac{(a_2 - b_2)(a_2 - c_2)}{(S_2)^2} U.$$

Again, the Lemma shows that

(4) $\quad \text{sgn}(B_1) = \text{sgn}(U).$

From (3) and (4), the theorem follows easily.

Q.E.D.

From the location of 1's saddle point, $b_1 \leqslant a_1 \leqslant c_1$. From the Lemma, $(d_2 - b_2)/(a_2 - c_2) > 0$. It follows that if 1's maximin strategy dominates, then Γ cannot be a B game, for $d_1 - c_1 \leqslant 0$ forces $U \leqslant 0$. The properties of Type II games under monotone transformations are given by the following corollary:

COROLLARY: Let Γ be a Type II game with players and strategies enumerated as in Theorem 2. Then Γ is a gen-B game iff Γ is an anti-B game with either $d_1 < b_1 = a_1 < c_1$ or $b_1 < a_1 < c_1 \geqslant d_1$.

Proof: Let Γ be a B game. Recall that $b_1 \leqslant a_1 \leqslant c_1$ and $S_2 > 0$. If f is any monotone transformation, then, by the Lemma, $S_2' > 0$. Now $d_1 > c_1$, for otherwise $\text{sgn}(U) \leqslant 0$ and Γ is not a B game. If $b_1 = a_1$, $x = 0$ dominates $x = 1$, contradicting the definition of Type II games. Therefore, to show that Γ is not a gen-B game, we can assume that $b_1 < a_1 \leqslant c_1 < d_1$. Choose any numbers a_1', b_1', c_1', and d_1' satisfying

$$d_1' - c_1' < (a_1' - b_1')\left(\frac{d_2 - b_2}{a_2 - c_2}\right)$$

and either $b_1' < a_1' = c_1' < d_1'$, if $a_1 = c_1$, or $b_1' < a_1' < c_1' < d_1'$, if $a_1 < c_1$. This choice can always be made. Let f_1 be chosen so that f_1 is a monotone increasing function mapping a_1, b_1, c_1, and d_1 to a_1', b_1', c_1', and d_1', respectively, and define $f = (f_1, \text{id})$, where id denotes the identity transformation. Then $f(\Gamma)$ is non-B, so Γ is not gen-B.

Now let Γ be an anti-B game. If $a_1 = b_1$, then $d_1 < c_1$ by Theorem 2. Furthermore, 1's saddle point is not unique unless $d_1 < b_1 = a_1 < c_1$. It is easy to check that such a game is gen-B. Now assume that $a_1 > b_1$. If $d_1 > c_1$, an argument similar to the one above shows that Γ is not gen-B. If $d_1 \leqslant c_1$, then $a_1 < c_1$, for otherwise 1's saddle point is not unique. In this case, $b_1 < a_1 < c_1 \geqslant d_1$, it is easy to check that Γ is gen-B.

Q.E.D.

Several observations follow from the corollary. A gen-B game of Type II is anti-B and always transforms into an anti-B game. The class of non-B games can be partitioned into three sub-classes, *viz.* games which can arise as images of B games, games which can arise as images of anti-B games, and games which can arise only as images of other non-B games. In the special case when the player with the saddle point has no equalities within his payoff matrix, the game is gen-B iff the maximin strategy of the player with the saddle point dominates, and the 'diagonal preferred' by the player without the saddle point does not contain the saddle point of his co-player. In comparison to Type I games, Type II games possess little 'Braithwaite-invariance' with respect to monotone transformations. Finally, we consider games of Type III.

THEOREM 3: Let Γ be a game of Type III, and define

$$T = (\underline{x} - \bar{x})(\bar{y} - \underline{y}).$$

Then Γ is a B game iff $\text{sgn}(S_1) = \text{sgn}(S_2) = \text{sgn}(T)$; Γ is an anti-B game iff $\text{sgn}(S_1) = \text{sgn}(S_2) = \text{sgn}(-T)$; and Γ is an imp-B game iff $\text{sgn}(T) = 0$.

Proof: As in Theorem 2, $\underline{x} = \alpha_1/S_1$, $\bar{x} = \beta_2/S_2$, $\underline{y} = \alpha_2/S_2$, and $\bar{y} = \beta_1/S_1$. Furthermore, for any x and y,

$$V_1(\underline{x}, y) = V_1(x, \bar{y}) = d_1 - \frac{\alpha_1 \beta_1}{S_1},$$

$$V_2(x, \underline{y}) = V_2(\bar{x}, y) = d_2 - \frac{\alpha_2 \beta_2}{S_2}.$$

From (1) and (2),

$$B_1 = \frac{\alpha_1 \beta_1}{S_1} - \frac{\alpha_1 \alpha_2 + \beta_1 \beta_2}{S_2} + \frac{\alpha_2 \beta_2 S_1}{(S_2)^2},$$

$$B_2 = \frac{\alpha_2 \beta_2}{S_2} - \frac{\alpha_1 \alpha_2 + \beta_1 \beta_2}{S_1} + \frac{\alpha_1 \beta_1 S_2}{(S_1)^2}.$$

It is easy to check that $B_1/S_1 = B_2/S_2$. This common value is

$$\frac{B_1}{S_1} = \frac{B_2}{S_2} = \left(\frac{\alpha_1}{S_1} - \frac{\beta_2}{S_2}\right)\left(\frac{\beta_1}{S_1} - \frac{\alpha_2}{S_2}\right) = T,$$

so that, for $i = 1$ and 2,

(5) $\quad \text{sgn}(B_i) = \text{sgn}(S_i) \cdot \text{sgn}(T)$.

Since $S_i = 0$ is impossible for Type III games, the theorem follows easily from (5).

Q.E.D.

By (5), a monotone transformation alters the primary class of a Type III game iff it alters $\text{sgn}(T)$. The next result shows that 'almost all' games of

Type III can exhibit a change of primary class under the action of a monotone transformation.

THEOREM 4: Let Γ be any game of Type III.

(a) If i can be chosen so that $a_i \neq d_i$, then a monotone transformation f exists making $T' > 0$. Furthermore, $f_{j(i)} = \text{id}$.

(b) If i can be chosen so that $b_i \neq c_i$, then a monotone transformation f exists making $T' < 0$. Furthermore, $f_{j(i)} = \text{id}$.

Proof: Observe that the interchange of the roles of the two players does not affect the value of T. Therefore $i = 1$ can be assumed in (a) and (b). Interchange of the strategies of any one player changes T to $-T$. Therefore, to prove (b), it is sufficient to prove (a). Finally, $a_1 < d_1$ can be assumed, since otherwise interchange of the strategies of both players makes this so, without affecting the value of T. Fix any $\varepsilon > 0$ satisfying

$$\varepsilon < \min[\bar{x}, \underline{y}, 1 - \bar{x}, 1 - \underline{y}].$$

If $S_1 > 0$, choose f_1 so that a_1', b_1', c_1', and d_1' satisfy

$$a_1' - \min(b_1', c_1') < \varepsilon(d_1' - \max(b_1', c_1'))$$

and let $f = (f_1, \text{id})$. Then $\bar{x}' = \bar{x}$ and $\underline{y}' = \underline{y}$. Now

$$a_1' - c_1' < \varepsilon(d_1' - b_1') < \frac{1 - (1 - \varepsilon)}{1 - \varepsilon}(d_1' - b_1')$$

$$\Rightarrow \alpha_1' = d_1' - b_1' > (1 - \varepsilon)(a_1' - b_1' - c_1' + d_1') = (1 - \varepsilon) S_1'$$

$$\Rightarrow \underline{x} > 1 - \varepsilon > \bar{x}.$$

By a similar argument

$$a_1' - b_1' < \varepsilon(d_1' - c_1')$$

$$\Rightarrow \bar{y}' > 1 - \varepsilon > \bar{y},$$

so that $T' > 0$. If $S_1 < 0$, choose f_1 so that

$$\max(b_1', c_1') - d_1' < \varepsilon(\min(b_1', c_1') - a_1').$$

Then the proof that $\underline{x}' < \varepsilon$ and $\bar{y}' < \varepsilon$ is analogous to the previous case.

Q.E.D.

One consequence of Theorem 4 is the identification of the gen-B games of Type III.

COROLLARY: If Γ is a game of Type III, then Γ is a gen-B game iff $\text{sgn}(S_1) = \text{sgn}(S_2)$, and for some i, $a_i \neq d_i$ or $b_i \neq c_i$.

Proof: If $a_i = d_i$ and $b_i = c_i$, for $i = 1$ and 2, then it is easy to check that $T' = 0$ for any monotone transformation f. Such a game is not gen-B, by Theorem 3. Otherwise, it is possible to find f so that $\text{sgn}(T') \neq 0$, by Theorem 4. The corollary now follows from (5).

Q.E.D.

From the Corollary and Theorem 3, any Type III game which is B or anti-B, is also gen-B. As well, some imp-B games are gen-B. Thus the class of gen-B games is larger than for Types I or II. But Theorem 4 shows that gen-B games of Type III can exhibit a change of primary class under monotone transformations; this is not true of gen-B games of Types I or II. Thus the 'Braithwaite-invariance' of Type III games is of a different character from that of Types I and II. If a game of Type III has no equalities within the payoff matrix of each player, then it is a gen-B game iff the 'diagonals preferred' by the two players coincide. Such a game, if it is gen-B, can be monotonically transformed into a B game, an anti-B game, and an imp-B game. Similarly, 'almost all' non-B games of Type III can be transformed into imp-B games. This concludes our investigation of Type III games.

Assume now that the payoffs of Γ are given not on interval scales, but on ordinal scales, and assume further that there are no equalities within the payoff matrix of each player. Then Γ is one of the 78 games of Rapoport and Guyer's taxonomy [3]. Since Γ has at most one saddle point for each player, its Type can be identified. Using the three Corollaries above, it can then be determined whether Γ is gen-B or not gen-B. Finally, if Γ is of Type I, the Corollary to Theorem 1 shows that the primary class of Γ can, in fact, be identified; then Γ is gen-B iff it is B or anti-B. This procedure has been applied to each game in the taxonomy, and the results are given in Table I.

We conclude with several remarks related to Table 1. Let Γ be a gen-B game (with payoffs on ordinal scales and no equalities within each payoff

TABLE I
Type and class of all games in Rapoport and Guyer's taxonomy

Type	Class	Games of Taxonomy	Count
I	B	66 ('Chicken')	1
(36 games)	anti-B	1, 2, 3, 4, 5, 6	6
	imp-B	7, 8, 9, 10, 11, 12 ('Prisoner's Dilemma'), 45, 46, 48, 61	10
	non-B	13, 14, 15, 16, 17, 18, 19, 20, 21, 22, 23, 26, 35, 36, 39, 49, 50, 55, 72	19
II	gen-B	24, 25, 27, 28, 29, 30, 31, 32, 33, 34, 37, 38	12
(32 games)	not gen-B	40, 41, 42, 43, 44, 47, 51, 52, 53, 54, 56, 57 58, 62, 65, 67, 70, 71, 77, 78	20
III	gen-B	59, 60, 63, 64 ('Two Musicians'), 68 ('Leader'), 69 ('Hero')	6
(10 games)	not gen-B	73, 74, 75, 76	4

matrix). All of the representations of Γ have the same primary class iff Γ is of Type I or II. In particular, every representation of Γ is a B game iff Γ is Game 66 ('Chicken'). Every representation of Γ is an anti-B game iff Γ is either one of the 6 anti-B games of Type I, or one of the 12 gen-B games of Type II. There exist two representations of Γ with different primary classes iff Γ is one of the 6 gen-B games of Type III; in this case, there exist representations of Γ as a B game, an anti-B game, and an imp-B game. The example given previously is Game 65, which is of Type II and not gen-B. Note that Rapoport and Guyer regard it as very similar to Braithwaite's 'Two Musicians' game, which is Game 64.

Dept. of Mathematics, University of Toronto

BIBLIOGRAPHY

[1] Braithwaite, R. B., *Theory of Games as a Tool for the Moral Philosopher*, Cambridge University Press, Cambridge, 1955.
[2] Luce, R. D. and Raiffa, H., *Games and Decisions*, John Wiley and Sons, New York, 1957.
[3] Rapoport, A. and Guyer, M., 'A Taxonomy of 2 × 2 Games', *General Systems Yearbook* **11** (1966) 203–214.

CLEMENT S. THOMAS

DESIGN AND CONDUCT OF METAGAME THEORETIC EXPERIMENTS

1. Introduction

In his pioneering treatise on metagame theory [3], Howard has presented a theory of stability of outcomes. Two further contributions to metagame theory have been a modified theory of stability by Hildebrand [1], and a theory on bargaining moves by Thomas [6]. Since all three are empirical scientific theories an important phase in the development of each has been that of experimental testing. The purpose of this paper is to report fully one of these experimental testing programmes and to outline the factors that must be considered in planning programmes of this sort.

We shall begin by a detailed description of the experimental programme designed and conducted to test the theory of bargaining behaviour reported in [5, 6]. This will be followed by a brief review of two prior experimental programmes that were conducted to test theories of stability of outcomes in a game [1, 2, 3]. The description in each case will consist of: a presentation of the theory; listing of testable assertions consequent to the theory; outline of the experimental procedure; and, results of the experiments. Using the experiments on bargaining behaviour for illustration we shall then give our comments on the design of metagame theoretic experiments. Throughout this paper we shall use the notation and terminology of Howard [4]. A brief resume of the basic ideas of metagame theory is given in Section 2.1, but the reader will have to refer to Howard [3, 4] for details.

Metagame experiments, unlike most other experiments in the social sciences, are completely deterministic in nature. The theory being tested asserts that a certain prediction will be confirmed in *all* cases: hence a single substantiated counterexample suffices to overthrow the theory. This is so unusual in the social sciences that we feel that our experiments are of a pioneering kind, and we shall describe them in this way. If we make little reference to the experience of other social scientists, this is because they are concerned with *probabilistic* experimentation.

2. Experiments for Thomas's Theory of Transitions

As noted, we propose to describe in detail the experimental verification of this theory of transitions. But first we give some basic concepts of metagame theory used in describing this theory and its consequences.

2.1. Basic Concepts from Metagame Theory

We start with the normal form of the game, $G = (S_1, S_2; M_1, M_2)$, where S_1 is the set of *strategies* for player 1 and M_1 is his preference function. A typical element of S_1 is written 's_1'. Then, unlike some game theorists, we define an 'outcome' as an ordered pair of strategies. Thus the set of outcomes is

$$S = S_1 \times S_2,$$

so that an outcome $\bar{s} \in S$ is an ordered pair

$$\bar{s} = (\bar{s}_1, \bar{s}_2).$$

We deal with players' preferences rather than utilities. Therefore we interpret $M_1 : S \to \mathcal{B}(S)$, where $\mathcal{B}(S)$ is the set of all non-null subsets of S, as follows:

$$M_1(\bar{s}) = \{s \mid s \text{ is not preferred to } \bar{s} \text{ by player 1}\}.$$

To simplify the notation we shall, in future, write $M_1(\bar{s})$ as $M_1\bar{s}$.

The set of outcomes *rational* for player 1 in G is

$$R_1(G) = \{\bar{s} \mid \forall s_1 : (s_1, \bar{s}_2) \in M_1\bar{s}\}.$$

For notational simplicity we sometimes write R_1 for $R_1(G)$.

The underlying idea of the metagame approach is that in order to analyse a game we must analyse the metagames based upon it. The definition of a 'general' metagame has been given by Howard [4] for the n-person case. We shall now give his definition, modified for the two-person case.

Let $\mathcal{K} = (K_1, K_2)$ be a pair such that for player 1 we have both

(i) $\quad K_1 \subseteq \{c_1 \mid c_1 \supset \emptyset; c_1 \subseteq S_1\} = \mathcal{B}(S_1),$

and

(ii) $\bigcup_{c_1 \in K_1} c_1 = S_1,$

and for player 2 we have both

(i) $K_2 \subseteq \{c_2 \mid c_2 \supset \emptyset; c_2 \subseteq S_2\} = \mathscr{B}(S_2),$

and

(ii) $\bigcup_{c_2 \in K_2} c_2 = S_2.$

Then, the first level metagame $\mathscr{K}G$ is defined to be a game

$$\mathscr{K}G = (X_1, X_2; M_1', M_2'),$$

where

$$X_1 = \{x_1 \mid x_1 = (f_1, c_1); \quad c_1 \in K_1; f_1 : K_2 \to c_1\},$$
$$X_2 = \{x_2 \mid x_2 = (f_2, c_2); \quad c_2 \in K_2; f_2 : K_1 \to c_2\},$$

and where for each $i \in \{1, 2\}$, M_i' is defined by

$$x \in M_i' \bar{x} \Leftrightarrow \beta x \in M_i \beta \bar{x};$$

in which expression β is an operator defined by

$$\beta(f_1, c_1; f_2, c_2) = (f_1(c_2), f_2(c_1)).$$

The metagame $\mathscr{K}G$ is a well-defined two person game. Proceeding in a manner similar to the above we may form a second-level metagame $\mathscr{K}'\mathscr{K}G$, based on $\mathscr{K}G$. Recursively we may obtain an rth-level metagame

$$\mathscr{K}_r \mathscr{K}_{(r-1)} \ldots \mathscr{K}_1 G,$$

where r is some non-negative integer. If $r=0$ we have the game, G, called the *basic* game. When $r \neq 0$ we have a *proper* metagame, which is a *descendant* of the basic game (and of itself). The set of all metagames $\mathscr{K}_r \mathscr{K}_{(r-1)} \ldots \mathscr{K}_1 G$ ($r=0, 1, 2, \ldots, \infty$) based on a given basic game G is called the *infinite metagame tree* based on G.

When we single out a particular metagame from the infinite tree, and do not wish to specify its level, we shall denote it by 'H'. A typical outcome in H is $x=(x_1, x_2)$; it is an ordered pair of metastrategies x_1, x_2. The set of all outcomes in H is denoted by $X(H)$, so that $x \in X(H)$.

A metagame outcome can be reduced to a basic outcome by a sufficient number of applications of the β operator (defined above); thus an outcome in $\mathscr{K}_r\mathscr{K}_{(r-1)}\ldots\mathscr{K}_1 G$ can be reduced by r successive applications of the β operator, i.e. β^r. When we do not wish to specify the level of metagame, we use the notation β^* which represents 'a sufficient number of applications of the β operator'. Thus if $x \in X(H)$ reduces to outcome s in the basic game, we say that $\beta^* x = s$.

The set of outcomes rational for player 1 in the metagame H is

$$R_1(H) = \{\bar{x} \mid \forall x_1 : (x_1, \bar{x}_2) \in M_1' \bar{x}\}.$$

The corresponding set of basic outcomes $\hat{R}_1(H)$, where $\hat{R}_1(H) = \beta^* R_1(H)$, is called the set of outcomes that are *metarational* for player 1 from the metagame H.

For players who know their opponent's preferences, in a two-person game, Howard [3] has defined the concept of inducement. An *inducement* is an action that a player takes which, if his opponent reacts rationally, will lead to a preferred outcome for himself. Howard thus asserts that a player will use his knowledge of his opponent's preferences to induce his opponent to benefit him by reacting rationally.

Consider, for example, the game in Figure 1. (Note that in this figure, as often in this paper, integer payoffs are used to stand for the values $M_i s$; they have merely ordinal significance and do not stand for quantities of anything.) In this game player 1 (row player) may move from (D, C) to (C, C) in order to induce player 2 (column player) to be rational and move to (C, D), which is preferred to (D, C) by player 1. Therefore, player 1 *induces* his best outcome at (C, C), and at (C, D). Likewise, player 2 induces his best outcome at (C, C) and (D, C).

	C	D
C	12	43
D	34	21

Fig. 1. A two-person 2×2 game.

In order to make most use of the concept of inducement we make one basic assumption: players' preferences are mutually ordinal. By this we mean that each player's preferences are ordinal, and that a player is indifferent between two outcomes only if both players are indifferent between those two outcomes. One can argue that this is also a very realistic assumption since real-life games are nearly always mutually ordinal. (See Howard [3] and Thomas [6] for fuller discussions of this point.)

Formally, a game $G(S_1, S_2; M_1, M_2)$ is *mutually ordinal* if and only if
(i) $\forall s, i: s \in M_i s$,
(ii) $\forall i: s \in \tilde{M}_i t, t \in \tilde{M}_i r \Rightarrow s \in \tilde{M}_i r$,
(iii) $\forall i: s \in M_i t, t \in M_i s \Rightarrow M_i s = M_i t$,
(iv) $(\exists i: M_i s \supset M_i t) \Rightarrow (\forall i: M_i s \neq M_i t)$.

Conditions (i) to (iii) define an ordinal game (that is, one in which players' preferences are ordinal). Fulfilment of condition (iv) makes the game mutually ordinal.

In a finite, two-person, mutually ordinal game, the *inducement value* for player 1 of outcome $\bar{s}$ is defined as $D_1(\bar{s}) = M_1 s$, where s is an outcome such that $s_1 = \bar{s}_1$, and $s \in R_2$. This value is unique. Similarly, in a proper metagame, H, the (unique) inducement value for player 1 of a metagame outcome $\bar{x}$ is given by $D_1(\bar{x}) = \beta^* M_1 x$, where x is a metagame outcome such that $x_1 = \bar{x}_1$, and $x \in R_2(H)$. For example, in Figure 1, $D_1(C, D) = 4$, $D_2(C, C) = 4$, and $D_2(D, D) = 3$.

This outlines all that we need to borrow from Howard [3, 4] for use in this paper. We shall next look at our theory of transitions.

2.2. *The Theory of Transitions*

Within the metagame theoretic framework this author has developed a theory [6] about players' moves in a finite two-person game from one outcome to another till stability is achieved. These moves, which we call *transitions*, could be mental. That is, they could be made in the course of thinking about the conflict that the game models, or, if the player is an organization of individuals, in the course of discussions within the organization. Alternatively, they could be stated during bargaining or negotiations. Otherwise, they could be positions taken up in a crisis, with the intention of threatening or forcing the opponent. In any case, they are completely reversible. We require this in order to distinguish between a transition and a final choice of strategy.

Before we state the theory let us give some definitions. In the basic game a *1-transition* is defined as a triplet $(s, s', 1)$ such that $s_2 = s'_2$. Similarly, a 1-transition in a metagame is defined as a triplet $(x, x', 1)$ such that $x_2 = x'_2$. We shall write the triplets $(s, s', 1)$ and $(x, x', 1)$ as $s \xrightarrow{1} s'$ and $x \xrightarrow{1} x'$ respectively.

The 1-transition $s \xrightarrow{1} s'$ is a move (in the basic game) which may be executed by player 1 and only by him. Similarly, $x \xrightarrow{1} x'$ is a move – in the metagame H such that $x, x' \in X(H)$ – which can be executed by player 1 and only by him.

If $s = s'$ (or in a metagame, $x = x'$) the transition $s \xrightarrow{1} s'$ (or $x \rightarrow x'$) is called an *identity* transition. It is said to occur when a player 'stays' at an outcome.

A 1-transition is called *rational* if and only if $s' \in R_1$. For example, in Figure 2, player 1 may make a non-identity rational transition from (D, C) to (C, C). Both players' identity transitions at (C, C) are also rational.

	C	D
C	23	42
D	14	31

Fig. 2. A game with rational transitions.

In two-person, finite mutually ordinal games with complete information we can define an *inducing transition* by a player as one whereby he induces an outcome that he prefers to the one he would have attained by being rational. Thus a 1-transition $s \xrightarrow{1} s'$ is *inducing* if and only if

$$D_1(s') \supset \max M_1(\langle s_2 \rangle),$$

where $\langle s_2 \rangle = \{\hat{s} \mid \hat{s}_2 = s_2\}$. For example, in Figure 1, $(D, C) \xrightarrow{1} (C, C)$ and $(C, D) \xrightarrow{1} (C, C)$ are inducing transitions made by players 1 and 2 respectively. The indentity transitions made by either player at (C, C) are likewise inducing.

If, as in Howard [3], we define a function 'M_i^j' which takes any one of player j's values into the corresponding value for player i, then the condition for $s \xrightarrow{1} s'$ to be inducing becomes

$$D_1(s') = M_1^2 \max M_2(\langle s_1' \rangle) \supset \max M_1(\langle s_2 \rangle) = M_1^2 D_2(s).$$

The condition for $x \xrightarrow{1} x'$ to be an inducing transition in the metagame H, where $x, x' \in X(H)$, may be written in an analogous manner and is left to the reader.

We can now state our theory of transitions. It asserts, firstly, that a player in the game G will perceive himself as playing in some metagame based on *G* or possibly in *G* itself. (This is the assumption of all 'metagame' theories.) Secondly, it asserts that players will always make either rational or inducing transitions in the metagame in which they perceive themselves.

Note that the first assertion does not imply that a player ever knows explicitly what metagame either he or his opponent is playing in. What we mean is that if it were possible to analyse a player's thought process we would find that it corresponds to a policy of his in some metagame *H*, and we could then say that he possibly perceives himself as being in *H*.

Even if we knew exactly which metagame a player was thinking in terms of we would not, in general, be able to analyse it because of its enormous size. Consequently, in order to apply our theory – and, especially, to test it – we need a *basic characterization* of a metagame transition in terms of outcomes in the basic game. Our approach to this problem of basic characterization is based on the concept of offered sets as we shall now see.

2.3. *Consequences of the Theory of Transitions*

Before we give any testable assertions consequent to our theory of transitions we shall characterize a metagame transition.

Consider the game in Figure 3. Suppose that the players are at outcome (C, D). Then, according to our theory, if player 1 confines his thinking to the basic game he will make only the rational 1-transition $(C, D) \to (D, D)$.

	C	D
C	33	14
D	41	22

Fig. 3. Prisoners' dilemma.

Next, suppose the players, still at (C, D), perceive themselves as being in the metagame $\mathcal{H}^2\mathcal{H}^1G$ (Figure 4). (This metagame, also written as 12*G*, has been cited frequently in the literature of metagame theory so we shall assume the reader is familiar with it. Briefly, it is the game which would result if player 2 chose his strategy in full knowledge of player 1's choice, and player 1 chose his strategy in knowledge of player 2's reaction to this prior information.) The fact that the players are at the basic outcome (C, D) now tells us very little; the players could be at any of the sixteen outcomes in the metagame which arise from the basic outcome (C, D), and we must know at which outcome they are if we are to predict what transitions from it are feasible.

We, therefore, see that it would not be enough for our purposes, to characterize an outcome in a metagame by its corresponding basic outcome. Furthermore, it would not do for us to characterize a transition in a metagame by a pair of basic outcomes corresponding to the metagame outcomes between which the transition occurs.

	C/C	D/D	C/D	D/C
C/C/C/C	33	14	33	14
D/D/D/D	41	22	22	41
D/D/D/C	41	22	22	14
D/D/C/D	41	22	33	41
D/D/C/C	41	22	33	14
D/C/D/D	41	14	22	41
D/C/D/C	41	14	22	14
D/C/C/D	41	14	33	41
D/C/C/C	41	14	33	14
C/D/D/D	33	22	22	41
C/D/D/C	33	22	22	14
C/D/C/D	33	22	33	41
C/D/C/C	33	22	33	14
C/C/D/D	33	14	22	41
C/C/D/C	33	14	22	14
C/C/C/D	33	14	33	41

Fig. 4. The metagame $\mathcal{H}^2\mathcal{H}^1G$ of Prisoners' Dilemma. $D/C/C/C$ represents the policy of D against C/C and C against the remaining three policies. C/D means 'choose C if player 1 chooses C and D if he chooses D'.

Instead, we characterize an outcome x in a metagame by a triplet (T_1, T_2, s) of two sets T_1, T_2, of basic outcomes, and a basic outcome, s, such that $T_1 = \beta^*\langle x_2 \rangle$, $T_2 = \beta^*\langle x_1 \rangle$, and $s = \beta^* x$.

This triplet is called a *metaoutcome*. Furthermore, we characterize a metagame transition $x \to x'$ by a pair of metaoutcomes corresponding to x and x'. Such a pair of metaoutcomes is called a *metatransition*.

For example, in Figure 4, the metaoutcome corresponding to $(D/D/D/D, C/D)$ is (T_1, T_2, s) where

$$T_1 = \{(C, C), (D, D)\},$$
$$T_2 = \{(D, C), (D, D)\},$$

and $s = (D, D)$. The metatransition corresponding to the transition $(D/D/D/D, C/D) \xrightarrow{1} (D/D/C/D, C/D)$ is $(T_1, T_2, s) \xrightarrow{1} (T_1, T_2^*, s^*)$ where T_1, T_2, s are as above, $T_2^* = \{(C, C), (D, C), (D, D)\}$, and $s^* = (C, C)$.

Notice that the fundamental concept used in the definition of a metaoutcome and a metatransition is that of a set of basic outcomes, T_1 or T_2. Such a set is called an *offered set*. It is the set of outcomes that a player 'offers' to his opponent by virtue of his choice of strategy or metastrategy. For instance, in Figure 2, if player 1 selects his strategy C he offers outcomes (C, C) and (C, D) to his opponent; we say that $T_2 = = \{(C, C), (C, D)\}$. The subscript '2' in T_2 denotes the player *to* whom the set is offered.

How can our theory be expressed in terms of offered sets, metaoutcomes, and metatransitions? Let us answer this question in terms of the metatransition $(T_1, T_2, s) \xrightarrow{1} (T_1, T_2^*, s^*)$.

First, let us note that since player 1 makes this metatransition (and the corresponding metagame transition) he may change the set he offers to his opponent but the set T_1 offered to him will not change. Hence, we have the same T_1 at both ends of the arrow.

If the transition is rational, then the end point will have the maximum value for player 1 of all the outcomes in T_1. That is

$$M_1 s^* = \max M_1(T_1).$$

If the transition is inducing then the value for player 1 of the outcome that is 'best' (in the sense that all outcomes are not preferred by him to this one) for player 2 in T_2^* will be greater than the value for 1 of his 'best' outcomes in T_1. That is, if, as in Section 2.1, we define a function 'M_1^2' that takes any one of player 2's values into the corresponding value for player 1, then we have

$$M_1^2 \max M_2(T_2^*) \supset \max M_1(T_1).$$

Let us illustrate this definition. First, look at the outcome (C, D) in

Figure 3. As we have explained before, $M_1(C, D) = 1$, and $M_2(C, D) = 4$. Then $M_1^2 M_2(C, D) = M_1(C, D) = 1$, and $M_2^1 M_1(C, D) = M_2(C, D) = 4$. Next, look at the example of a metatransition that we give below; the reader may refer to Figure 3 for the values, and Figure 4 for the corresponding transition, which is $(D/C/D/C, D/D) \xrightarrow{1} (C/D/C/D, D/D)$.

$$T_1 = \{(C, D), (D, D)\},$$
$$T_2 = \{(C, D), (D, C), (D, D)\},$$
$$s = (C, D),$$
$$T_2^* = \{(C, C), (D, C), (D, D)\},$$
$$s^* = (D, D).$$

Here $\max M_1(T_1) = M_1(D, D) = 2 = M_1 s^*$. Hence, it is a rational metatransition. Also, since

$$M_1^2 \max M_2(T_2^*) = M_1(C, C) = 3,$$

which is greater than $\max M_1(T_1) = 2$, the metatransition is also inducing.

We can now state some testable consequences of our theory of transitions. Stated in terms of metatransitions our theory asserts that $(T_1, T_2, s) \to (T_1, T_2^*, s^*)$ is a possible metatransition if and only if either

$$M_1 s^* = \max M_1(T_1)$$

or

$$M_1^2 \max M_2(T_2^*) \supset \max M_1(T_1).$$

In other words, we assert that a player, upon receiving an offer, will pursue at least one of the following two courses of action:

(a) he will select the best outcome for himself in it.
(b) he will make a counteroffer such that his opponent's best outcome in it is also better for himself than the outcome he would have obtained by maximising as in (a).

The method of testing our theory is now clear: make experimental subjects exchange offers and see whether our assertions regarding their behaviour are valid. However, before we can proceed with experimentation a few more questions must be answered.

Firstly, what constitutes a valid offer? We have already given an informal definition of an offered set. Formally an offered set to player 1 is defined as a set T_1 of basic outcomes such that

$$\exists H, \bar{x}_2 \in X_2(H); T_1 = \beta^* \langle \bar{x}_2 \rangle.$$

The family of offered sets to player 1 from the metagame H is denoted by $\hat{\mathscr{I}}_1(H)$. It can be shown [5, 6] that if H' is a descendant of H, then $\hat{\mathscr{I}}_1(H) \subseteq \hat{\mathscr{I}}_1(H')$.

We now define the family of *general* offered sets to player 1:

$$\mathscr{G}_1(G) = \bigcup_H \hat{\mathscr{I}}_1(H).$$

Every member of this family is an offered set to player 1 from some metagame. This is thus the family of all valid offers to player 1.

We have characterized this family (see [5] and [6]) as follows:

$$T_1 \in \mathscr{G}_1(G) \Leftrightarrow \forall s_1 \exists s_2 : s \in T_1.$$

Thus a set of outcomes is an offered set by a player from some metagame if and only if it contains at least one outcome from every strategy choice of his opponent.

We also need a characterization for metaoutcomes. Formally, a *metaoutcome* is defined as a triplet (T_1, T_2, s) such that

$$\exists H, x \in X(H): \begin{cases} T_1 = \beta^* \langle x_2 \rangle; \\ T_2 = \beta^* \langle x_1 \rangle; \\ s = \beta^* x. \end{cases}$$

The family of metaoutcomes from the metagame H is denoted by $\mathscr{Q}(H)$.

We define the family of general metaoutcomes $\mathscr{Q}^g(G)$ as follows:

$$\mathscr{Q}^g(G) = \bigcup_H \mathscr{Q}(H).$$

This family contains every triplet (T_1, T_2, s) that is a metaoutcome from some metagame. Its characterization has been given by us [6] as follows:

$$(T_1, T_2, s) \in \mathscr{Q}^g(G) \Leftrightarrow T_1 \in \mathscr{G}_1(G), \quad T_2 \in \mathscr{G}_2(G), \quad s \in T_1 \cap T_2.$$

Thus any pair of intersecting valid offers – one for each player – can constitute a metaoutcome. There is a theorem [6] which states that a metaoutcome from a metagame H is also valid from every descendant of H.

Finally, we need a characterization for metatransitions. Formally, we define a *1-metatransition* as a pair of metaoutcomes $(T_1, T_2, s) \xrightarrow{1} (T_1, T_2^*, s^*)$ such that

$$\exists H, (x, (x_1^*, x_2)) \in X(H)): \begin{cases} T_1 = \beta^* \langle x_2 \rangle; \\ T_2 = \beta^* \langle x_1 \rangle; \\ T_2^* = \beta^* \langle x_1^* \rangle; \\ s = \beta^* x; \\ s^* = \beta^* (x_1^*, x_2). \end{cases}$$

Again we have shown [6] that a 1-metatransition from a metagame H is also a metatransition from any and every descendant of H.

We have also shown [6] that $(T_1, T_2, s) \xrightarrow{1} (T_1, T_2^*, s^*)$ is a metatransition if and only if $T_1 \in \mathscr{G}_1(G)$; $T_2, T_2^* \in \mathscr{G}_2(G)$; $s \in T_1 \cap T_2$; $s^* \in T_1 \cap T_2^*$. Thus any valid pair of metaoutcomes having the same offered set T_1 can constitute a 1-metatransition.

With these results we are ready to perform the experimental verification of our assertions. We shall now describe our experimental procedure.

2.4. *Experimental Procedure*

Subjects in these experiments were university students. Each 'player' consisted of two students and there were four players – W, X, Y, and Z – every evening. The author and an assistant were the instructor-referees for the games. Players W and X were met and instructed by one instructor, and players Y and Z by the other in another room, and care was taken to keep the two pairs of players separate. The procedure for the evening consisted of (i) instruction of the players, (ii) practice of the games, and (iii) playing the actual games.

During the instruction players were told that they would practise against each other (i.e. W against X, and Y against Z) or against the instructor, but that the actual two games would be played against two different players whom they would not see, hear or otherwise know the identity of before, during, or after the games. W and X were to be row players in both the actual games, and Y and Z column players. They were also told that the actual games would be similar in all respects to the practice games except that the monetary values (not the preference order) of the outcomes would be different. Their earnings for the evening would consist of their earnings from each of the two actual games; for each person this would be one half of the value of the final outcome of the game, this final outcome being arrived at in the manner described during the instruction period. Their objective should be to make money.

After each instructor had instructed his players and made them practise the first game, the instructors switched rooms so as to conduct each other's players in the second practice game. This was done to help minimize errors in instruction and to better evaluate whether the players were indeed familiar with the experimental procedure.

When all four players were ready to play the actual games they were separated into four rooms and the games begun. The instructors then refereed the games as follows:

	Referee I	Referee II
Game 1	WY	XZ
Game 2	WZ	XY

(Here WY represents the game in which W was row player and Y was column player.)

Thus, two games were played every evening, each game being played by two different pairs of players, giving four trials. The games used were 3×3 games. An example is given in Figure 5.

	C	D	E
R	33	74	42
S	29	91	57
T	65	18	86

Fig. 5. A two-person 3×3 game.

The experimental procedure was as follows: Each player (consisting of two subjects as noted above) was given a game matrix showing the value of each outcome to him and to his opponent. His value was given in dollars and cents, while his opponent's was given as a letter – 'a' being his opponent's best, followed by 'b', and so on. An example is given in Figure 6 which shows player W's sheet in Game 15, where his opponent was Y. This is the game of Figure 5. (The corresponding practice game would have been numbered 15P, and would have been the same except for the dollar amounts, which would have been in the same order but of different magnitudes.) Figure 7 shows player *Y*'s sheet in the same game. All information pertaining to row player was always given in blue, while information pertaining to column player was always given in red on the game sheet. Players were told that their opponent's payoffs were not necessarily of corresponding magnitudes to their own. (For instance, if a player's highest payoff was $10, his opponent's highest payoff would not necessarily be $10.)

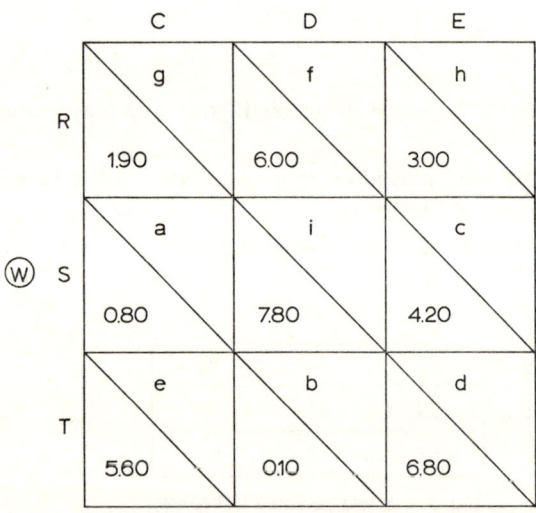

Fig. 6. Game matrix for player *W*.

When they were ready to proceed each player was asked to predict his opponent's strategy choice and then state his own strategy choice. The pair composed of each player's stated strategy choice determined the starting outcome of the game, which was then communicated to each player by the referee. The starting player (randomly determined beforehand but not revealed to the players till now) was then asked to offer a set of outcomes to his opponent, including at least the following:

(i)   the starting outcome, and,

(ii) one outcome from every strategy choice of his opponent (that is, player 1 had to offer at least one $s \in \langle s_2 \rangle$, for every $s_2 \in S_2$).

This offer was made by placing the player's letter in the appropriate cells of the matrix on the offered set form (an example is shown in Figure 8). The referee then took the offered set form to the opponent who was asked to select an outcome out of this offered set, by circling his opponent's letter in the appropriate cell. He could then end the game,

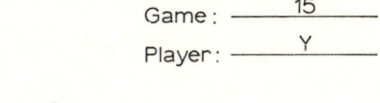

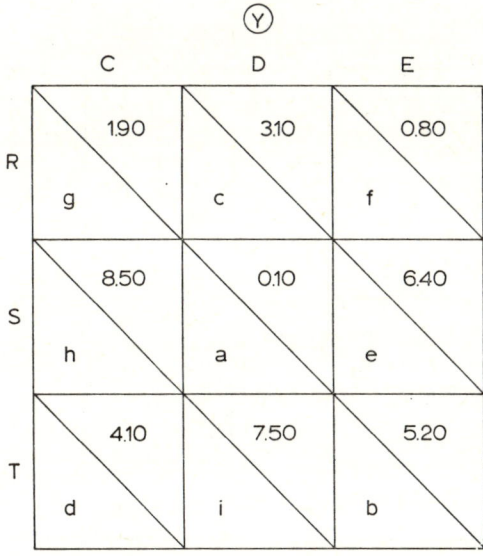

Fig. 7. Game matrix for player Y.

in which case his circled outcome would be the final outcome (payoff point) of the game. If he wished to continue he would offer a set to his opponent (including, at least, the circled outcome and one outcome from every strategy choice of his opponent) provided that the referee did not announce that the game was over. For, each game had a fixed maximum duration in terms of number of moves. This number was predetermined (according to a truncated normal distribution about a mean of 5 moves) but was not revealed to the players.

The game thus proceeded till either a player decided to end the game or the predetermined number of moves were up, whichever occurred first. In the first few games of this programme a player was allowed to end the game at the starting outcome when he received the first offered set. Later this was changed so that a player could end the game upon receiving the first offered set at any outcome but the starting outcome. The reason we made this change was that the starting outcome could not

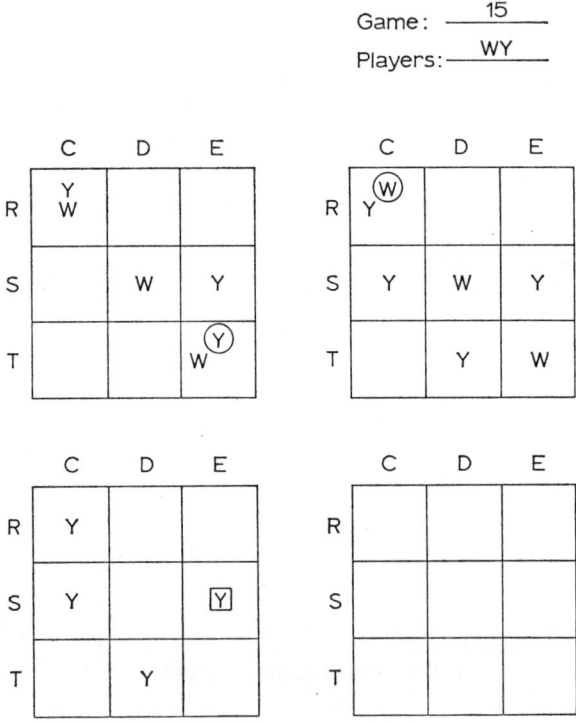

Fig. 8. Offered set form.

always be taken as a legitimate part of an offer by the starting player; the rules forced him to offer this outcome whether he wanted to or not. Consequently, the old rule curtailed the liberty of action of the starting player while giving his opponent the advantage of ending the bargaining before the starting player could exercise any controls. Note that this change of rules did not curtail the opponent's liberty of action; he could still try to end the bargaining at the starting point if he wished by maintaining his original strategy choice. After exchange of the first set of offers, of course, either player could end the game after circling any outcome he wished.

Thus, any outcome that a player offered to his opponent could possibly be the final outcome if the opponent selected it, either by the opponent's ending the game, or by the number of moves running out. Also, any outcome that a player selected (circled) from his opponent's offered set could be the final one even if he did not wish it to be, for the game could end upon his selecting it, or else his opponent could select it (for he would have to offer it) and make it the final outcome (either by choice or by the moves running out once again).

If, upon receiving an offered set, a player did not wish to select an outcome, i.e. the set was totally unattractive to him, he could request a restart of the game. In this case, the referee would notify the opponent and each would perform a one-shot procedure (as in the beginning) thus determining an outcome. This outcome then took the place of the outcome the player would normally have selected (circled). However, he could not choose to end the game at this outcome. Thus, if the duration of the game was not complete, the referee would ask the player for an offered set including this restart outcome and the game would continue. Furthermore, this restart outcome did not enjoy the protection that was afforded the starting outcome; the opponent *could* end the game after selecting this restart outcome. We were at first apprehensive about allowing players this restart option, lest through it the experiment degenerate into a series of one-shot games. We were relieved, then, to find that players used this option advisedly for they saw that they were giving up some of their freedom in using it.

There was one more procedural step: whenever a player selected an outcome and then offered a set, he was required to fill a move questionnaire, an example of which is shown in Figure 9. This questionnaire was designed to elicit responses to questions that the player would normally resolve while making his selection of an outcome and then offering a set. This written record of those responses assisted the player in making an educated move, while enabling the referee to ascertain whether the player had indeed made the move that he thought he was making.

Note that the questionnaire did not 'guide' the players or attempt to

GAME: _____15_____

Player: _____Y_____

1. We have selected outcome _____RC_____.

2. Among outcomes offered to *us* by our opponents, the outcome that is best for us is _____TE_____.

3. Among outcomes that we now offer to our opponents, the outcome that is best for our opponents is _____SE_____.

4. Now compare all the outcomes in the game with the outcome you have named in Question 3.
   Outcomes that are better for both us *and* our opponents are: (please check outcomes in matrix)

	C	D	E
R			
S			
T			

   none

Fig. 9. Move questionnaire.

persuade them to obey any theory. The questions simply brought out certain facts about the game and the players' moves – facts in the light of which the players had to make decisions, concerning which our theory made some assertions. The questionnaire thus brought out these facts but did not suggest to the players as to what to do about them.

This, then, was the experimental procedure. Figures 6–9 show an example of a game. Players W and Y were given the game matrices of Figures 6 and 7. W chose row T, while Y chose column E, so that TE was the starting outcome. The players were then told that Y was the starting player, and he then offered the set $\{RC, SE, TE\}$ of outcomes on an offered set form (Figure 8). This form was then taken to player W who selected outcome TE. This being the starting outcome and the first offer, W could not choose to end the game here, so since the predetermined number of moves was not over, W offered the set $\{RC, SD, TE\}$. He indicated this by placing his letter (W) in the appropriate cells in the first matrix in Figure 8, and repeating them in the next free matrix. W

also filled a move questionnaire (not shown) similar to the one in Figure 9. The referee checked this questionnaire and then left it with W. The offered set form was then taken to Y, who selected outcome RC, circling it on the form, and opted against ending the game. The referee bade the two subjects who comprised player Y to continue, so they offered $\{RC, SC, SE, TD\}$, which they indicated by placing a Y in those cells in the second matrix and the next free matrix of Figure 8. They also filled the move questionnaire of Figure 9. The offered set form was then taken to W, who selected SE and decided to end the game (indicated by a square around the Y in cell SE of Figure 8). Thus, the two subjects comprising player W earned $2.10 each from this game, while their opponents (Y) earned $3.20 each.

As the reader can see, our experimental procedure provides a test of our theory because of the fact that the sets offered by the players in our experimental procedure are stated explicitly, instead of remaining hidden in the minds or motives of the players as would have been the case in a one-shot or reaction game procedure (see Section 3 below), or even if players were made to bargain across a table.

2.5. *Results*

Forty-seven experiments were run in all, using 20 different 3×3 games. In these experiments, 133 metatransitions were effected, thus affording us 133 tests of our theory that players will make only rational or inducing metatransitions.

62 of these metatransitions were rational and the remaining 71 were inducing. Our theory was thus validated in each of these tests without a single counterexample.

As we have mentioned before our theory could not have been tested had we resorted to a one-shot or reaction game procedure. Such procedures are, however, useful for testing theories of stability. In the next section we shall briefly discuss how they were used for just such a purpose.

3. EXPERIMENTS FOR HOWARD'S AND HILDEBRAND'S THEORIES OF STABILITY

Howard's theory of metarationality [2, 3, 4] states that a player will always be rational in the metagame in which he regards himself as being.

That is, if a player thinks in terms of his opponent's policies and his own counter policies in a way that is modelled by a metagame H then, given a policy of his opponent, he will select only that counterpolicy which yields a rational outcome for him in the metagame H. It follows that a basic outcome will be stable for a player only if it is metarational for him.

We use the term 'stable' here to mean anticipated in some way. Note that the player may not necessarily desire the game to end at the anticipated outcome; he simply predicts that the game will, in fact, end there. In a one-shot game a stable outcome for a player may be operationally defined as an outcome that results when he is asked to first predict his opponent's strategy choice and then choose his own strategy.

Subjects in Howard's experiments consisted again of university students. Each player (consisting of one subject) knew his payoffs in dollars. As to opponents' payoffs, various conditions were employed. In some games each player had full information (including the dollar amount) about his opponent's payoffs. In others he had knowledge of the ordering of his opponent's preferences. In still others, he had no information or misinformation about his opponent's preferences. Opponents in a game never saw or otherwise knew the identity of each other. In some games no communication was allowed between players. In others, written messages were allowed. The messages were censored so as to exclude reference to anything which might take place between the two subjects either before or after the experiment. Also, the subjects were not allowed to reveal any information about their payoffs that their opponents did not already have. The messages were sent back and forth on a single sheet of paper till both players were satisfied.

The play of the game consisted of a one-shot procedure. That is, each player wrote down his prediction of his opponent's strategy choice, and then wrote his own choice. The outcome determined by the pair of players' own choices was the final result of the game, and each player was paid the dollar value to him of that outcome.

Howard's assertion that a player's anticipated outcome will be metarational for him was confirmed in 172 trials, with not a single counterexample. The assertion that a player's anticipated outcome will be rational for him was falsified in 34 out of 172 trials, thus confirming, conclusively, that the hypothesis that players will always be rational in the basic game is false.

Howard's theory of metarationality is a theory of stability of outcomes; it asserts that if an outcome is stable for a player it must be metarational for him. Hildebrand [1] has modified this theory by adding a further necessary condition for stability, based on the concept of a *perfect optimum set*. We shall not give either the definition or the characterization of it, but shall merely state that a perfect optimum set, P, is a set of basic outcomes such that neither player ever has any rational or inducing motivation to move out of it, regardless of the metagame in which he perceives himself. The set of perfect optimum sets is denoted by $\bar{\pi}$ in Hildebrand [1].

Hildebrand's theory of stability states that an outcome can be stable for a player only if it is metarational for him and if no perfect optimum set exists in which all outcomes are preferred by him to it. Thus, while Howard's theory of metarationality asserts that

$$\bar{s} \text{ is stable for player } 1 \Rightarrow \bar{s} \in \bigcup_H \hat{R}_1(H),$$

Hildebrand's theory of stability asserts that

$$\bar{s} \text{ is stable for player } 1 \Rightarrow \begin{cases} \bar{s} \in \bigcup \hat{R}_1(H); \\ \forall (P \in \bar{\pi}) \, \exists (s \in P) : s \in M_1 \bar{s}. \end{cases}$$

This theory was tested using a reaction game procedure as we shall describe shortly. For such a procedure Howard [3] has operationally defined a stable outcome for a player as one at which he is willing to be paid off before the time allotted for the experiment has elapsed. The implication is that an outcome is stable for a player if he feels that he cannot reasonably expect to improve upon it for himself.

Subjects in these experiments were university students. The games used were 3×3 games. Each 'player' consisted of two subjects. Opponents in an actual game never knew each others' identities. No communication (in the form of messages) was allowed between players.

As in Thomas's and Howard's experiments, the procedure for the evening consisted of: (i) instruction of subjects; (ii) playing practice games; and (iii) playing the actual games. In all three cases, too, players never practised with subjects against whom they played the actual games. Also, players never played against the same opponents more than once.

In Hildebrand's experiments, printed instruction sheets (supplemented by verbal instruction) were tried but were later discarded in favour of fully verbal instruction. The reaction game procedure was as follows:

Each player was given a game matrix, similar to the ones described in Section 2.4 above (see Figures 6 and 7), and told which the (predetermined) starting outcome was. Each player then filled a questionnaire, indicating in it whether or not he was willing to 'accept' the outcome as the final outcome of the game. This acceptance/rejection question was put to each player before and after each move. Whenever both players 'accepted' an outcome the game was terminated at that outcome. (Acceptance is, thus, taken to be synonymous with stability for a player.) Players moved alternately, beginning with the (predetermined) starting player. A move consisted of a choice of strategy. Thus, row player, in his turn, could move (or stay) within the column that he found himself in, and column player could change columns while remaining in the row in which he found himself. Following each move the player filled a move questionnaire.

The game ended when both players accepted an outcome, or else when the ending rule terminated it. Various ending rules were tried, the final choice being a predetermined number of moves. Players were not told this number of moves, though under some rules they were informed when their move was to be the last.

Sixty-two experiments were run in all, using 23 different 3×3 games. This afforded 207 tests of the theory concerning stability of outcomes for players. The theory was validated in 206 of these tests; there was but one apparent contradiction.

With this we end our discussion of the three experimental programmes that have been conducted to date to test metagame theories. While the report of the first programme (due to Thomas) is detailed, the reader will have to refer to the original references for details of the procedure if he wishes to replicate the experiments of the other two programmes.

Although the experimental procedure differed in the three programmes, they were, nevertheless, designed according to the same underlying principles. These principles, which we shall now enumerate, are of relevance to designs of all metagame theoretic experiments and must, we feel, be adhered to if the experiments are to serve their intended purpose.

4. Design of deterministic metagame experiments

We shall illustrate the discussion in this section by examples from our experimental programme described in detail in Section 2 above. In all other respects what we say here applies to all metagame theoretic experiments and probably also to experiments that are conducted to test other deterministic empirical assertions of game theory.

The need for experiments to test metagame theoretic assertions is the same as the need for testing empirical scientific assertions in general – although we should stress again that we are concerned with deterministic experiments. Metagame theory deals with real-life conflict situations. Consequently, its assertions must be applicable in real-life. However, before the metagame theorist applies his theory to a real-life situation he would like to make reasonably sure that it is not false, and for this assurance he conducts experiments under controlled conditions. Once his theory, or some modified form of it, is corroborated by experimental evidence he may probationally accept it as correct (till some valid counterexample has been cited) and proceed to apply it to real-life situations.

Note that a metagame theory, like any scientific theory, can never be proved correct; all we can ever say is that thus far no valid contradiction to the theory has been found. The theory is thus accepted as tentatively correct and applied accordingly, till a valid counterexample necessitates its modification or discard.

The metagame theorist's main reason for performing experiments, then, is to test the assertions of his theory and obtain a legitimate falsification of it if possible.

Sometimes, deterministic experiments are run solely as heuristic aids in developing a theory. Such heuristic experiments are not intended to test any assertions but are solely to give guidance as to what assertions might be profitably included in a proposed theory.

Whether experiments are run to test a theory or to aid in developing a theory we feel that they must be designed and conducted according to certain principles if they are to fulfil their intended purpose. These principles, however, are in some ways different from those which apply to probabilistic experiments. We believe that they should be as follows:

The first step in designing deterministic experiments is to find a procedure that evokes the behaviour about which assertions are being made and tested. For instance, our assertions were about the offered sets that players exchange while bargaining or negotiating. Consequently, we had to set up a procedure wherein players could exchange offered sets which were stated explicitly instead of remaining hidden in the minds or motives of the players as would have been the case in a one-shot or reaction game procedure, or if players were made to bargain across a table. In order that each offered set T_i be valid (that is, $T_i \in \mathcal{G}_i$) players had to be made to select the sets such that

$$\forall s_i \; \exists s_j : s \in T_i.$$

Recall that this is the characterization of the family $\mathcal{G}_i$ of general offered sets to player i. Furthermore, the procedure had to be designed such that players made metatransitions between metaoutcomes, in accordance with the characterizations we have given in Section 2.3 above.

The correct model of the play of a given game in normal form is afforded by a one-shot procedure (von Neumann and Morgenstern [7]), wherein players predict each other's strategy choices and then state their own choices. Any other procedure must have a normal form equivalent in some sense to this if it is to be regarded as giving evidence concerning the given normal form game. This equivalence becomes difficult to show if the procedure is complicated. For our experiments we have shown [6] that certain basic structures are the same as the corresponding structures of the given normal form game.

Having developed the outline of the experimental procedure, the next step is to fill in the details. The various forms or data sheets that are to be used must be designed for unambiguity and facility of use, especially for the experimental subjects but also for the referee and experimenter. In experiments dealing with human behaviour, especially when the procedure is relatively complicated such as ours, it is imperative that no effort be spared towards this end, even down to such details as colour-coding of the game matrices (in our experiments we used blue and red for row and column players respectively) and identifying players by letters.

In a real-world conflict the game analyst obtains the game model (that is, the strategies of the various players, and players' preferences over the various outcomes) by querying the player or decision-maker for whom the conflict analysis is being effected. In other words, the game analyst works with the game that is perceived by the decision-maker. In the laboratory, however, the experimenter supplies the game and must ensure that the players understand it and that they have the preferences that he wishes them to have.

Understanding the game means knowing what each player's possible choices of strategies are, and what possible counterpolicies the opponent has a choice of in response to one's own policies. In other words, it means being familiar with the game matrix when it is given in normal form. To ensure that players in experiments understand the game it should be kept simple and players should be made to practise playing it a number of times till both they and the experimenter (or whoever instructs the players) are satisfied on that count. The practice games should be similar to the actual games in all respects but for the monetary payoff magnitudes. Also, players should never practise with anyone who will be an opponent in the actual games; this is to prevent the effects of learning and experience in game-playing from contaminating their preferences in unknown amounts. For this reason, too, players should never play against the same opponents in more than one actual game.

To ensure that players have the preferences assigned to the various outcomes by the experimenter, these outcomes should be assigned monetary payoffs in the appropriate order of magnitude. Furthermore, these payoffs should be of sufficient magnitude to be meaningful to the subjects. For this reason it is useful to employ 'poor' university students as subjects. Subjects should be told that their objective should be to make as much money for themselves as they can.

Subject's preferences can be contaminated in many ways, so various steps should be taken to diminish this source of experimental error. To avoid preference contamination due to altruism and chivalry subjects should be isolated from their opponents throughout the experiment – as well as before and after it – so that opponents never know each others' identities in any way. They should be informed of these arrangements prior to playing the games.

Another possible source of preference contamination is some subjects' penchant for fairness and equality (especially when it does not require too great a sacrifice on their part). In a real-life conflict this will have been incorporated into a player's preferences before he gives them to the game analyst, but in the laboratory there is no way of doing this. To alleviate it one should not give players their opponents' monetary payoff values; in our experiments we simply gave opponents' preferences as letters – 'a' being his best, followed by 'b', and so on.

One source of experimental error, and possible preference contamination, is subjects' being pressed for time or rushed in any way. Real-world decision makers do not have unlimited time at their disposal, and in any case laboratory experiments must have limitations on their length. Consequently, if the procedure calls for a number of moves, there should be a limit on the number of moves, rather than the time, allotted to each game. Such a limit renders delaying tactics ineffectual and allows players to think through their moves more fully as indeed they should. Players should not be told the number of moves allotted for a game; therefore, each move has to be a serious, deliberate one as it could be the last.

Finally, steps should be taken to ensure that players have understood and remember the rules of the game. The practice games facilitate this. Sometimes it is helpful to provide a sheet of instructions in addition to – or instead of – the verbal instruction session. We feel however, that for more complicated procedures, such as ours, written instructions are more a hindrance to understanding than an aid. Consequently, in our experiments we depended solely upon verbal instruction. However, we switched instructors half way through the instruction period, the better to evaluate whether the subjects were ready to play the actual games, and also to minimise instructional errors. As an added safeguard we asked quite an extensive series of questions of our subjects during these instruction-practice sessions. In the actual games players filled a move questionnaire (Figure 9) each time they moved. The questions in this form helped the referee and the subjects to tell whether the latter were playing with proper understanding of the game and its rules.

Unfortunately, while as noted above the use of verbal instruction curbs the occurrence of invalid counterexamples, it can also inhibit the occurrence of any counterexample at all. In probabilistic experiments this could entirely invalidate the results. However, in deterministic experiments its effect is merely to weaken the power of the experiments. One way of alleviating its effect is by switching instructors half way through the instruction period, as was done in our experiments.

To facilitate in understanding the game and the rules of the game it is useful to have each 'player' consisting of two (or more) subjects. Partners can then discuss each decision and help one another in overcoming errors and oversights. This also diminishes preference contamination. For, a subject must first convince his partner(s) before any change in a player's preferences can occur. Moreover, the referee becomes aware of any potential or actual preference contamination as he listens to the discussion among the subjects comprising a 'player', and he can, therefore, decide whether the experimental results are valid. He can also ascertain the cause of contamination and remedy it in future experiments.

Care must be exercised to guard against inadvertently 'coaching' or advising subjects as to what to do, during the instruction, practice and actual experimental sessions. Instructions must be explicit and strictly enforced; anything that can be classified as a suggestion or advice – to be followed at the subject's discretion – is probably out of place anyway and is best omitted. This author also feels that any instructions or communication from the instructor or referee to the subjects should always be the truth, and the subjects should be advised of this fact from the beginning. For, any information that the subjects should not have can be very easily denied them without resorting to bluffing.

We must point out, once again, that these experiments are not statistical in nature. We do not seek a probabilistic estimate of the validity of our theory. Our aim, rather, is to obtain a legitimate falsification of the theory, if such a falsification can be found. For, as we have pointed out, a determinate empirical scientific theory, such as a metagame theory, can never be proved correct, or even estimated to be correct within certain confidence limits. One legitimate falsification is enough to invalidate a scientific theory, so we must strive to obtain at least one such falsification if we can. Of course, the greater the number of experiments that are conducted, the better are the chances of finding a valid falsification to one's theory, if such a falsification exists. Therefore, the experimental programme should include as many experiments as possible within the constraints of time and finances. But, in these experiments, subjects need not be chosen randomly. On the one hand, persons who we know will not heed the instructions or take the experiments seriously should be screened out. On the other hand, if a certain type of person seems more likely to behave in legitimate contradiction to the theory under test, the experimenter might use more subjects of that type.

5. Acknowledgements

The research reported in Section 2 and the latter half of Section 3 was funded by grants from the National Research Council of Canada.

All three experimental programmes were conducted under the guidance of Dr Nigel Howard. Experiments reported in the latter half of Section 3 were conducted by Dr Nigel Howard, Mr Philip Hildebrand, Mr Ian Shepanik, Dr R. K. Ragade, and this author.

I am grateful to Dr Nigel Howard for his guidance and advice in the preparation of this paper.

Ministry of Transport
Ottawa, Ontario, Canada

BIBLIOGRAPHY

[1] Hildebrand, P., 'Stable Outcomes in Two-Person Games', Ph.D. Thesis, University of Waterloo, 1973.
[2] Howard, N., 'The Theory of Metagames', Ph.D. Thesis, University of London, 1968.
[3] Howard, N., *Paradoxes of Rationality: Theory of Metagames and Political Behaviour*, MIT Press, Cambridge, Mass., 1971.
[4] Howard, N., ''General' Metagames: An Extension of the Metagame Concept', this volume, p. 261.
[5] Thomas, C. S., 'The Theory of Transitions in Two-Person Metagames', paper presented at the Midwest Regional Conference, International Studies Association and Peace Research Society (International), Toronto, May, 1972.
[6] Thomas, C. S., 'Transitions and Offered Sets in Two-Person Games', Ph.D. Thesis, University of Waterloo, 1973.
[7] Von Neumann, J. and Morgenstern, O., *Theory of Games and Economic Behaviour* (3rd ed.), Princeton University Press, Princeton, N.J., 1953.

ANATOL RAPOPORT AND J. PERNER

TESTING NASH'S SOLUTION OF THE COOPERATIVE GAME

Typically, the outcomes of a two-person non-constantsum game can be partitioned into two classes – Pareto-optimal and non-Pareto-optimal. The former jointly dominate the latter in the sense that both players prefer any Pareto-optimal outcome to any non-Pareto-optimal one. However, in order to achieve a Pareto-optimal outcome, the strategy choices of the two players must, in general, be coordinated. Moreover, a Pareto-optimal outcome may not be an equilibrium; so that it may be in the interest of each player to 'move away' from it, even though if *both* move away, both may suffer an impairment of payoffs.

A two-person non-constantsum game is played cooperatively if the players are able to coordinate their strategy choices and to make binding agreements about which strategies they will choose. Consequently, in such a game a Pareto-optimal outcome can always be achieved. In general, however, there will be more than one Pareto-optimal outcome, as well as a continuum of such outcomes spanned by mixtures of pairs of them. If so, the preferences of the players for different Pareto-optimal outcomes will be opposed. The problem posed in the theory of the cooperative game is that of singling out from the set of Pareto-optimal outcomes preferably a unique one that would constitute the 'solution' of the cooperative game, i.e., the resolution of the conflict between the players which ensues after they have cooperated to insure that the outcome is one of the Pareto-optimal ones.

Several solutions have been proposed. [One, due to Braithwaite (1955) is discussed in another paper in this volume (M. Kilgour, 'On 2 × 2 Games and Braithwaite's Arbitration Scheme.')] Any such solution must take into account the relative bargaining advantages that the players have vis-à-vis each other. In the spirit of game theory, these advantages must derive from the structure of the game itself, not from, say, the players' individual bargaining skills. The bargaining advantages of the respective players are reflected in the position of some reference point in the payoff space – the plane with the respective players' payoffs as coordinates –

with respect to the Pareto-optimal set. For instance, in Braithwaite's solution, this reference point is taken as the intersection of the 'security levels' of the two players – the minimum payoffs that each could guarantee himself by a proper choice of strategy. As we shall see in a moment, this choice of reference point does not fully represent the bargaining leverage of each player.

Consider the game represented by Matrix 1.

	Column S_2	Column T_2
Row S_1	2 / 0	0 / 2
Row T_1	−1 / −2	−2 / −1

Matrix 1

By choosing S_1, Row can guarantee himself 0. But this choice does not exert any leverage against Column. For if Row points out to Column his ability to get 0, Column can well reply, 'All right, play so as to get your 0, that is, choose S_1. This suits me fine, for I shall choose S_2 and get the largest possible payoff'.

Intuitively, Row feels that he should get more than 0. In fact, if the two could coordinate their strategies so as to effect outcomes $S_1 S_2$ and $S_1 T_2$, each with probability $\frac{1}{2}$ (or alternate between the two in iterated plays), Row would get on the average 2.5 per play, as would Column. This 'solution' may seem eminently fair to Row, because of the apparent symmetry of the payoff matrix. Nevertheless the symmetry is only apparent. Note that strategy S dominates strategy T for both players. Therefore in a non-cooperative game the outcome $S_1 S_2$ is the only 'defensible' one (being the only equilibrium); and this outcome favors Column. On the other hand, Row is not helpless. For Row's 'cooperation' (choice of S_1) is necessary if Column is to get a non-negative payoff (let alone 2). Row, in other words, has a 'threat'. He can threaten to choose T_1, in which case both players will get negative payoffs. On the basis of this 'power to strike', as it were, Row can hope to induce Column to share a portion of the largest payoff with him, either by choosing S_2 and T_2 with certain positive probabilities (thus insuring a positive expected payoff to Row) or in an iterated game by alternating with certain relative frequencies between S_2 and T_2. The question is what is the frequency or probability with which Column should choose T_2 (while Row 'cooperates' by choosing S_1) that reflects the bargaining leverages of the players vis-a-vis each other.

A solution that takes due account of the 'threats' that the players can exert against each other was proposed by Nash (1953). Consider first the

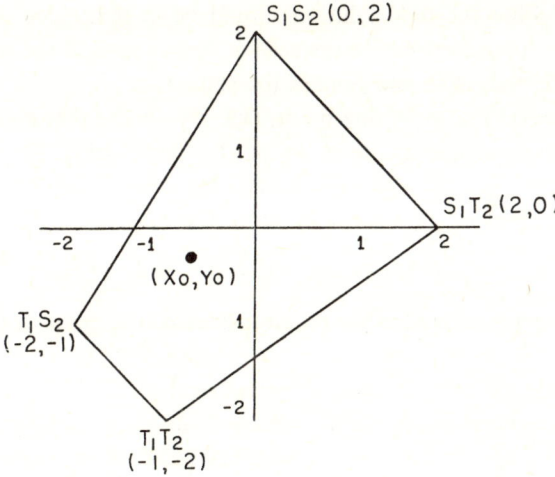

Fig. 1.

situation where the 'threats' are imposed by an outside authority. In other words, if the two players fail to agree on which point of the Pareto-optimal set shall be the outcome of the game, they get a pair of payoffs fixed in advance. For instance, in the above game, let it be known that if the players cannot agree on a mixture of the Pareto-optimal outcomes S_1S_2 and S_1T_2, then Row will get $x_0 < 0$, and Column will get $y_0 < 0$. The situation is shown in Figure 1.

With this 'threat' outcome as a reference point, the 'proper' mixture of S_1S_2 and S_1T_2 can be calculated as a consequence of four apparently reasonable axioms, proposed by Nash.

AXIOM 1. Symmetry. (The solution should be invariant with respect to the re-labeling of the players.)

AXIOM 2. Pareto-optimality. (The solution should be on the Pareto-optimal set.)

AXIOM 3. Invariance with respect to independent positive linear transformations of the payoffs. (This is in consequence of the fact that in two-person game theory, the payoffs are usually assumed to be given only on an interval scale.)

AXIOM 4. Invariance with respect to irrelevant alternatives. (If a solution of a game has been determined and if the payoff space is extended or curtailed, while the threat point remains the same, then, in the case of extension, either the new solution should be in the added region or it should remain unchanged; in the case of curtailment, if the solution has not been deleted, it should remain the same.)

Nash showed that only one point, (x^*, y^*), on the Pareto-optimal set satisfies all four axioms. It is found by maximizing the expression $(x-x_0)(y(x)-y_0)$ where $y(x)$ is defined by the equation of the Pareto-optimal set. That is, x^* is the solution of the equation

(1) $\qquad y(x) + y'(x)(x - x_0) - y_0 = 0$, and $y^* = y(x^*)$.

The next step is to abandon the requirement that the threat point be given in advance. Nash showed how it can be chosen by the players themselves as an intersection of their optimal 'threat strategies'.

Since, in general $y'(x)<0$, it appears from equation (1) that it is to Row's advantage to have x_0 as large as possible and y_0 as small as possible and that Column's preferences are opposite. In effect, therefore, in choosing their threat strategies to determine the point (x_0, y_0), the players are playing a game, in which their interests are diametrically opposed. In a 2×2 game, this auxiliary game can be easily determined as follows. To fix ideas, let the feasible payoff space of the game be a quadrilateral such as is shown in Figure 1, where the 'northeast' edge is the Pareto-optimal set. (More general situations require slight modifications of the procedure.) Normalize the payoff scales so that this line segment has slope -1. (This is already done in the game being considered.) Form a game in which the payoffs are the algebraic differences of the respective players' payoffs in the normalized game. The resulting difference game is shown in Matrix 2. Clearly, it is a zero-sum game.

	Column	
	S_2	T_2
S_1	2 / −2	−2 / 2
T_1	1 / −1	−1 / 1

Row

Matrix 2

The solution of the zero-sum game determines the threat outcome of the original game. In our example, the zero-sum game has a saddle point at $T_1 S_2$. Thus, the outcome $T_1 S_2$ of the original game is the threat point, and the solution of that game is calculated with reference to it. Noting that on the Pareto-optimal line $y = 2 - x$, $y'(x) = -1$, and that the payoffs at $T_1 S_2$ are $x_0 = -2$; $y_0 = -1$, we substitute these values into (1), solve

(2) $\quad (2 - x) - (x + 2) + 1 = 0$,

and obtain $x^* = \frac{1}{2}$, $y^* = \frac{3}{2}$, the solution sought. It can be implemented in iterated play if Column chooses S_2 with frequency 0.75 and T_2 with frequency 0.25, while Row always chooses S_1.

In view of the reasonableness of Nash's axioms and because the method seems to capture the essential pressures that the two bargainers in a cooperative game can exert on one another, it seems natural to expect that the solution deduced in this manner will somehow reflect the 'settlement' at which 'rational' players will arrive.

One question, to be sure, remains. Do the payoff numbers in the game matrix reflect the 'real' utilities of the players? To raise this question means to become concerned with the problem of assessing the players' 'utility functions' for money or for whatever the payoffs represent. While methods for assessing these functions experimentally have been proposed, the results are anything but conclusive. That is to say, it is by no means possible to get definite consistent answers to the questions posed; so a diversion into an altogether different experimental area purportedly to 'refine' the formulation of a game by determining the players' 'real' utilities, in order to test Nash's theory experimentally, does not seem to be warranted. A much simpler way out is to put the proposed solution of the cooperative game to experimental test directly by assuming that the payoffs represent utilities. Whether the theory is then corroborated or refuted will not constitute a final verdict (because of our ignorance of the 'real' utilities.) Assuming, however, that no verdict is likely to be final in a test of a behavioral theory, we should proceed on the simplifying assumption that in a 2×2 game played for trivial stakes, the utilities of the payoffs are positive linear functions of the payoff entries. Since the solution is invariant with respect to positive linear transformations of the payoffs (see Axiom 3 above), we may consider these payoffs as representing utilities.

The design of an experiment to test Nash's solution presents some problems. Clearly, we do not wish to restrict our subjects to persons familiar with the concepts of game theory, such as mixed strategies, expected payoffs, etc. Yet these concepts are incorporated in the way the subjects theoretically choose their strategies to determine the threat point in the general case and to determine the solution. With regard to the choice of the threat outcome, mixed strategies can be dispensed with by choosing game matrices where the threat outcome is an intersection of pure strategies. However the 'mixed' nature of the solution is an essential feature of the theory and so cannot be dispensed with.

We have adopted the following procedure. After the rules of a 2×2 game in matrix form have been explained to the subjects, they are instructed to play in two phases. In phase 1, each chooses a strategy independently with the understanding that the resulting outcome will go in effect each time they fail to agree on the outcome to be chosen in phase 2. It is pointed out to the subjects that in choosing a strategy in phase 1, it is by no means always advantageous to try to maximize one's own payoff in the outcome. The resulting outcome does not award the payoffs associated with it but only serves as a 'threat', which goes into effect in case of a deadlock in phase 2. Thus, it is to each player's advantage to have the other's payoff as low as possible in the threat outcome in order to weaken his bargaining leverage in phase 2. On the other hand, making the other's payoff low in the threat outcome may be possible only if one's own payoff is also low or even lower. The advantage of one must be balanced against the disadvantage of the other.

Having determined the threat outcome (either the one theoretically prescribed or some other one), the players go on to phase 2 (bargaining). Now they are allowed to discuss the game and to come to some agreement on how to make subsequent choices. The choices made in phase 2 are not strategies (as in a non-cooperative game) but outcomes. They are made as follows. One player proposes one of the four outcomes. The other can agree or not. If he agrees, the outcome obtains on that play of the game. If not, for example if he insists on another outcome, the threat outcome determined in phase 1 goes into effect.

The players take turns in proposing the outcome for ten consecutive plays. This procedure seemed to us preferable to taking turns on each successive play, because the latter suggests an 'easy compromise', whereby each player agrees to the other's proposal and 'tit-for-tat' serves as control over the agreement. By prolonging the sequence to ten plays, we discourage this solution, for in that case, a player, by agreeing to the other's proposals must 'trust' him for a longer time that he will reciprocate later. This, we felt, would make the trivial alternating agreement solution less salient.

The players were also given the option after each sequence of ten plays to revise the threat outcome chosen in phase 1. This, too, was done to preclude trivial solutions. Consider, once again, the game represented by Matrix 1. Suppose $S_1 S_2$ instead of $T_1 S_2$ is chosen as the threat outcome. Then Row is 'helpless' vis-a-vis column. Column can always propose $S_1 S_2$, where he gets 2 and Row gets 0. Whether Row agrees or not, $S_1 S_2$ obtains. On the other hand, if Row proposes $S_1 T_2$, when it is his turn, column can always disagree. Then outcome $S_1 S_2$ obtains as the threat outcome.

It is, of course, interesting psychologically to see whether under these circumstances, the player who has complete control over the situation will utilize his advantage to the hilt. This can be done by examining the results in phase 2 when such 'wrong' threat points are chosen. However, we shall not report those results here. It turned out that when a threat outcome was chosen that conferred all power on one player, it was always revised. The players settled on the 'correct' threat outcome after at most 20 plays in phase 2, after which they played 100 iterations.

The four games used in our experiment are shown in Matrices 3–6. The identifying numbers are those used in a previously published taxonomy of 2×2 games (Rapoport and Guyer, 1966).

	S_2		T_2	
S_1		5		0
	0		5	
T_1		−1		−2
	−2		−1	

Matrix 3
Game #19

	S_2		T_2	
S_1		6		0
	0		10	
T_1		−3		−4
	−1		1	

Matrix 4
Game #21

	S_2		T_2	
S_1		10		2
	2		10	
T_1		3		7
	−5		7	

Matrix 5
Game #44

	S_2		T_2	
S_1		12		−12
	3		−4	
T_1		−3		3
	−9		12	

Matrix 6
Game #64

TABLE I
Payoffs in Game #19

Pair	Row	Column	
1	1.91	2.93	+
2	1.80	3.20	+
3	2.50	2.50	0
4	−0.53	−1.01	−
5	1.75	2.50	+
6	0.06	3.00	+
7	2.43	2.49	+
8	2.43	2.49	+
9	0.99	2.17	+
10	2.50	2.50	0
Mean	1.58	2.27	+
Solution	2.00	3.00	

The results, obtained from ten pairs of male students at the University of Toronto, are shown in Tables I–IV. The first two columns show the mean payoffs accruing to Row in 100–120 iterations. The entry in the last column is +, if the top dog received the larger payoff, −, if the underdog received the larger payoff, 0, if both received the same payoff. The top dog in each game is defined as the player to whom the Nash solution awards the larger payoff. He is Row in Game 21 and Column in the other three games. The last row of each table shows the Nash solution of each game for comparison.

The Nash solution of Game #19 awards 2 to Row and 3 to Column. We observe that Column gets more than Row in seven of the 10 pairs. Two pairs shared equally and one (Pair 4) is anomalous. Clearly, the negative payoffs received by that pair show that there have been many deadlocks, and that in cases of agreement Row, the underdog, was able to have his way more frequently. The observed discrepancy in Column's favor is smaller than prescribed, but is probably significant.

The Nash solution of Game #21 awards 7 to Row and 1.8 to Column. Clearly, in this game the theory is not corroborated. The mean payoffs of the two players are nearly equal. That the underdog gets more than the top dog in five cases, however, is an artifact, a consequence of the way the players alternated between $S_1 S_2$ and $S_1 T_2$. Namely, Column happened to get his largest payoff in the first sequence of ten plays and also in the last of 11 sequences. In pairs 4 and 6, this situation was reversed. Actually, therefore, all pairs except pair 1 agreed to share equally in this game.

It is interesting to see why Row is theoretically the top dog in this game. The Nash solution can be easily determined graphically by drawing a line

TABLE II
Payoffs in Game #21

Pair	Row	Column	
1	4.89	2.97	+
2	3.75	3.75	0
3	3.60	3.84	−
4	4.00	3.55	+
5	3.75	3.75	0
6	4.00	3.55	+
7	3.60	3.90	−
8	3.60	3.84	−
9	3.60	3.84	−
10	3.60	3.84	−
Mean	3.84	3.64	+
Solution	7.00	1.80	

from the threat point $(-1, -3)$ to the Pareto-optimal line, with a slope which is the negative of the slope of the latter. If follows that Row is favored if the slopes are numerically small. On the other hand, Column is favored if he stands to lose less at the threat point than Row. In game 21, the two effects are opposite. However the slope effect is the stronger. Thus, the numerical payoff to Row is larger. We conjecture that the relative magnitudes of these two effects were not appreciated by the players. The effects simply canceled out, suggesting equal division.

TABLE III

Payoffs in Game #44

Pair	Row	Column	
1	1.64	8.19	+
2	4.56	7.72	+
3	4.62	7.67	+
4	5.86	6.90	+
5	3.86	8.79	+
6	2.82	7.84	+
7	0.77	6.37	+
8	3.25	9.25	+
9	2.58	6.85	+
10	4.67	8.40	+
Mean	3.46	7.80	+
Solution	4.33	8.60	

Nash's solution awards 4.33 to Row and 8.60 to Column in Game #44. Here the discrepancy between the underdog's and the top dog's payoffs is quite clear. There are no instances of equal division, and the discrepancy in the mean payoffs (4.34) is quite close to that prescribed by the theory (4.27). Referring to the game matrix, we observe the following. Equal division could be effected either by alternating between S_1S_2 and S_1T_2 or by repeating T_1T_2. Of these two alternatives, the latter seems more attractive, since it awards each player 7, whereas the former awards only 6 on the average. However, in Nash's theory, the *sum* of the payoffs of the two players is not defined, inasmuch as the payoffs are considered to be given on an interval scale, invariant with respect to positive linear transformations. Therefore a 'compromise' cannot be effected by splitting the joint payoff but only by fixing the probabilities of choice of the two

NASH'S SOLUTION OF COOPERATIVE GAME   113

relevant outcomes (or, their frequencies in iterated play). Because of the position of the threat outcome, Column is strongly favored in this game. And this circumstance must have considerable salience for the players, because Row suffers a loss in case of deadlock, whereas Column gets a positive payoff. In fact, this payoff is actually larger than what Column gets in S_1T_2. Therefore Column has a strong argument. He need *never* consent to an occurrence of S_1T_2, since he gets more in the threat outcome than in S_1T_2, which is Row's most preferred outcome. Column can also hold out for more than 7, which is what equal division via repeated T_1T_2 awards him. He can, therefore, insist on some alternation between S_1S_2 and T_1T_2. The fact that Column gets more than 7 in seven of the runs attests to the strength of this argument.

TABLE IV
Payoffs in Game #64

Pair	Row	Column	
1	7.50	7.50	0
2	2.73	5.00	+
3	5.43	7.12	+
4	7.50	7.50	0
5	9.30	4.59	−
6	7.50	7.50	0
7	7.50	7.50	0
8	7.05	7.95	+
9	5.42	8.74	+
10	5.61	7.11	+
Mean	6.55	7.05	+
Solution	4.50	10.50	

In Game #64 the Nash solution awards 4.50 to Row and 10.50 to Column. Since the slope of the Pareto-optimal line, connecting (3, 12) and (12, 3) is −1, the difference is due entirely to the disparity of payoffs at the threat point. Apparently, however, although the disparity (6 in Column's favor) is comparable to that in the previous game (8), it is not nearly as salient in the minds of the players as the former, probably because both players get negative payoffs in case of deadlock, which was not the case in Game #44. From Table IV we see that although Column's payoff is larger than Row's, the difference does not compare with that prescribed by the theory. In fact, four of the ten pairs shared equally (an easy compromise effected by alternating between S_1S_2 and T_1T_2), and in one case (Pair 5), Row actually got more.

In summary, we see that many factors influence the outcome of a cooperative two-person game, perturbing the theoretical result, which might occur 'in vacuo', as it were. Naturally, the individual personalities of the players, their bargaining abilities, their relations to each other, etc. must all play a part in these perturbations, factors that are naturally of interest to social psychologists. However, if we adhere to the spirit of game theory, we must disregard these factors, for the theory is concerned not with the idiosyncracies of the players but only with the structure of the game. The players are simply assumed to be 'rational'. Their 'rationality' is supposed to impart to them an appreciation of the structural features of the game and so lead them to the solution, which must be accepted as 'fair' if the axioms from which it is derived are seen as reasonable ones.

We attribute the discrepancies between the theoretical solutions and the observed results not to individual characteristics of the players but to their common perceptions, which are, apparently, at variance with at least one of the axioms. In particular, Axiom 3 implies that the utilities of the two players are not comparable (an assumption often invoked in some theories of economics.) But it seems that the players *do* compare the numerical magnitudes of their payoffs, which accounts for the salience of the equal payoff outcomes. Further, Axiom 3 precludes a distinction between positive and negative payoffs: the ones can be turned into the others by a positive linear transformation. However the different results observed in Games #44 and #64 suggest that whether the payoffs at the threat point are of the same or of opposite sign makes a considerable difference in the bargaining position of the players.

It seems instructive, therefore, to use the formal theory of the cooperative game not as a model to be 'corroborated' or 'refuted' but rather as an anchorage for a behavioral theory – a base line, to which to refer the discrepancies observed in game behavior with the view of constructing a more psychologically oriented theory of conflict and of conflict resolution.

BIBLIOGRAPHY

Braithwaite, R. B., *Theory of Games as a Tool for the Moral Philosopher*, Cambridge University Press, Cambridge, 1955.

Nash, J. F., 'Two-Person Cooperative Games', *Econometrica* **21** (1953) 128–140.

Rapoport, A. and Guyer, M., 'A Taxonomy of 2×2 Games', *General Systems* **11** (1966) 203–214.

PART II

N-PERSON GAMES

JAMES P. KAHAN* AND AMNON RAPOPORT

TEST OF THE BARGAINING SET AND KERNEL MODELS IN THREE-PERSON GAMES**

ABSTRACT. Twelve groups of undergraduate subjects participated in computer-controlled experiments designed to test the bargaining set and kernel models for n-person games in characteristic function form. Each group played four iterations each of 5 three-person games in which $v(A) = v(B) = v(C) = v(ABC) = 0$, and $v(AB) > v(AC) > v(BC) > 0$. The effects of (i) the communication rules governing the negotiations, (ii) the differences among the payoffs assigned to each coalition, and (iii) learning were systematically investigated.

The results are analyzed in terms of the frequencies of different coalition types that were formed, the disbursements of the payoffs, and the characterization of the bargaining process. They show the predominance of coalition AB, support the bargaining set and kernel models as predictors of final outcomes, reveal significant effects due to the latter two independent variables but not the former, and provide useful information about the nature of the bargaining process.

I. INTRODUCTION

The study of coalition formation is a topic of long standing among social psychologists. Observational interest dates at least from Georg Simmel (1902), and experimental evidence dates from the mid-1950s (Vinacke and Arkoff, 1957). Most of the experiments on coalition formation in the triad have followed the Pachisi board format described in Caplow (1968). In this game format, the three players each move counters around a Pachisi board at a rate determined by a chance mechanism in conjunction with experimenter-imposed multiplicative weights. The first player to reach the goal-square on the board is the winner. If two or more players form a coalition, their counters are combined and move as one counter with a multiplicative weight equal to the sum of the weights of the members of the coalition. Each player signals which of the other players he would like to enter into coalition with, and, if there is a mutual choice, a coalition is formed. Allocation of the rewards (typically a fixed reward no matter which person or coalition wins the game) is decided upon by the members of a coalition after it has tentatively formed, in a manner often not reported explicitly (*e.g.*, Vinacke *et al.*, 1966).

It is the contention of the present paper that the experimental paradigm described above is inadequate for the study of coalition formation for two major reasons. First, the paradigm is limited to games with a fixed reward for each coalition. As such, it is insensitive to the fact that different coalitions in real-life may well command different amounts of desirable resources. A decision as to which coalition to join is typically based on such considerations. In political considerations, for example, more than mere majority power gives a ruling coalition power; there is also the question of the legitimacy of the government. In democratic forms of rule, the greater the proportion of citizens represented in the government, the greater the legitimacy of that government. In game-theoretic terms, then, our first objection is that most situations modelled by games are inherently non-zero-sum games. Second, and more importantly, the paradigm strictly separates the process of coalition formation from the division of rewards. Clearly, when the very reason for forming a coalition is to increase one's gains, one will consider the amount of gains that may be won from different coalitions before deciding which coalition to join. Negotiations may occur with different parties simultaneously, and offers may be carefully weighed one against the other, before final decisions are made. This bargaining process is entirely neglected in the Pachisi paradigm. Some recent work (Komorita and Chertkoff, 1973) shows awareness of the problem of negotiations affecting disbursements of payoffs, but even here coalition choice and outcome determination are separated.

The experimental paradigm presented below is designed to overcome these two inadequacies. Thus, coalition formation is scrutinized through examination of negotiated divisions of the rewards jointly available to the members of the coalition. This paradigm necessarily involves somewhat complicated experimental procedures, which were until recently not available within psychological laboratories, and have only become possible with the recent development of the digital computer as an experimental instrument. Prior to such instrumentation, the study of coalition formation through negotiations had only been undertaken in a rigidly narrowed scope (Riker, 1967; Selten and Schuster, 1968) or as a year long project without the opportunity for experimental control of psychologically interesting variables or systematic observation of the bargaining process (Maschler, 1965).

The orientation of our experimental paradigm arises from mathematical game theory. Familiarity with the nature, purpose, and principal ideas of game theory in general (see, *e.g.*, Luce and Raiffa, 1957; Rapoport, 1966) is assumed. We begin by a brief review of the mathematical background of *n*-person games, which is less known (but see, *e.g.*, Rapoport, 1970), and continue with a description of the experimental task.

A. *Theories of n-Person Games*

The major mathematical approach to *n*-person games considers them from the point of view of the *characteristic function*. Briefly described, an *n*-person game in characteristic function form is defined by naming all of the possible nonempty subsets of the *n* players in the game, and assigning a real number value $v(X)$ to each subset so named. The value represents a measure of the utility jointly commanded by that subset in coalition against all of the other players in the game. The term 'coalition' is used in a neutral sense, without institutional or structural implications. The assigned real-valued set function v, the number and identification of the *n* players, and the rules governing communication among them completely define the game. A solution to the game is a specification of which coalitions should form, and how the players in each formed coalition should divide among themselves the utility that they jointly control.

The present experiment is designed to investigate a class of characteristic function three-person games which have the special property that $v(A) = v(B) = v(C) = v(ABC) = 0$, and $v(AB)$, $v(AC)$, and $v(BC)$ are all positive but not necessarily equal to one another. Such games may be termed *quota games*, because quotas ω_i may be assigned to each player such that $v(ij) = \omega_i + \omega_j$ for each of the three possible coalitions AB, AC, and BC, where $i, j = A, B, C$, and $i \neq j$.

Unlike the case of two-person, zero-sum games, where the minimax principle provides a single compelling solution, there exists a multiplicity of solutions of *n*-person games in characteristic function form. Presently, at least five theories for the *n*-person cooperative game may be considered, namely, von Neumann and Morgenstern's *solution* (1947), Shapley's *value* (1953), Aumann and Maschler's *bargaining set* (1964), Davis and Maschler's *kernel* (1965), and Horowitz's *competitive bargaining set* (1973). For the quota games investigated in the present paper, only the bargaining set and kernel theories are of immediate interest. The solution of von Neumann and Morgenstern is unsatisfactory for our purposes as there are an infinite number of solution sets to the quota game. Because $v(ABC)=0$ in the quota games presented here, the Shapley value does not provide a solution. And in the particular case where $n=3$, Horowitz's competitive bargaining set theory reduces to the bargaining set theory.

For the particular case of three-person quota games, the two remaining theories – the bargaining set and the kernel – yield the same solution, although from different assumptions. We shall briefly present the principal ideas of both theories, and then show how they relate to the present experiment.

B. *The Bargaining Set*

Consider a cooperative n-person game in characteristic function form, described by the ordered pair $(v; N)$, where $N = \{1, 2, ..., n\}$ is a set of n players and v is the characteristic function of the game. The nonempty subsets of N are called *coalitions*; they form the domain of v. $v(X)$ is assumed to satisfy two properties:

(i) $v(X) \geq 0$ for each coalition X,
(ii) $v\{i\} = 0$ for each one-person coalition.

Let X be an m-partition of N, satisfying

$$X_j \cap X_k = \emptyset, \quad \text{if } j \neq k, \quad \text{and} \quad \bigcup_{j=1}^{m} X_j = N.$$

An *outcome* of the game is represented by a *payoff configuration* (*p.c.*).

$$(\mathbf{x}; \mathsf{X}) = (x_1, x_2, ..., x_n; X_1, X_2, ..., X_m),$$

where $\mathbf{x} = (x_1, x_2, ..., x_n)$ is the vector of utilities gained by each player called the *payoff vector*), and $\mathsf{X} = (X_1, X_2, ..., X_m)$ represents the proposed *coalition structure* (*c.s.*).

It is assumed that the members of each coalition $X_j \in \mathsf{X}$ are able to and do make use of all of the utility that they can jointly command, i.e.,

$$\sum_{i \in X_j} x_i = v(X_j), \quad j = 1, 2, ..., m.$$

Furthermore, it is assumed that a rational player does not ask for less than what he could obtain if he were not to participate in any coalition.

Thus, a *p.c.* is assumed to satisfy *individual rationality*, i.e.,

$$x_i \geq 0, \quad i = 1, 2, \ldots, n.$$

A third, key assumption is that every pair of players who are members of the same coalition have to reach a stable outcome representing their strengths and weaknesses. The distinctiveness of the bargaining set theory is how this stability is defined, through the concepts of *objection* and *counter-objection*. Before formally defining these terms, we shall demonstrate them through an example.

Consider a three-person quota game with the characteristic function $v(AB)=95$, $v(AC)=90$, $v(BC)=65$, and $v(A)=v(B)=v(C)=v(ABC)=0$. Suppose that players A and B have formed a coalition, agreeing on an 80:15 split. The *p.c.* so described is

(80, 15, 0; *AB*, *C*).

Player A has a threat, or an objection against B. He may point out that he can offer 5 points to player C (which are preferred to the zero points C is getting in the present situation), thus obtaining 85 points (as $v(AC)=90$), which is more than the 80 he is now slated to receive. Player A can take this action without the consent of B. Similarly, B has an objection against A. He can offer 15 points to C via coalition BC, reserving 50 points for himself. Of these two objections, the latter is *justified*, but the former is not. Justification rests upon whether or not the player objected against has available a counter to the objection. Thus, A's objection against B is unjustified, because B can make a counter-offer, giving C 50 points and keeping 15 for himself. C would prefer the 50 points to the 5 offered him by A, and B maintains his 15 points that he had in the original AB coalition. On the other hand, B's objection against A is justified, because there is no way that A can offer C 15 or more points and still maintain his original 80 points.

A coalition is defined as stable when neither member of the coalition has a justifiable objection against the other. For the example presented above, and excluding the case where no coalition is formed, this holds true for the following three *p.c.*'s only:

(60, 35, 0; *AB*, *C*),
(60, 0, 30; *AC*, *B*),
(0, 35, 30; *A*, *BC*).

The definitions of objection and counter-objection may now be formally presented. Let $(\mathbf{x}; \mathsf{X})$ be a *p.c.* and let k and l be two members of a coalition X_j, $X_j \in \mathsf{X}$. For a coalition Y and a payoff vector for its members, the pair $(\mathbf{y}; \mathsf{Y})$ is called an *objection* of k against l in $(\mathbf{x}; \mathsf{X})$, if

(i) $k, l \in X_j, k \in Y, l \notin Y$,

(ii) $\sum_{i \in Y} y_i = v(Y)$,

(iii) $y_k > x_k, y_i \geqslant x_i$ for all $i \in Y$.

In his objection $(\mathbf{y}; \mathsf{Y})$, player k claims that he can gain more in the new coalition Y, without the consent of l, and that this situation is reasonable because the new partners of k in Y get at least what they got in $(\mathbf{x}; \mathsf{X})$. Players which are not included either in the original coalition or in the objection are not considered at all, but for three-person games all players are necessarily considered.

For a coalition Z and a payoff vector $\mathbf{z}$, the pair $(\mathbf{z}; \mathsf{Z})$ is called a *counter-objection* to the objection $(\mathbf{y}; \mathsf{Y})$, if

(i) $l \in Z, k \notin Z$,

(ii) $\sum_{i \in Z} z_i = v(Z)$,

(iii) $z_i \geqslant x_i$ for all $i \in Z$,

(iv) $z_i \geqslant y_i$ for all $i \in Z \cap Y$.

In lodging a counter-objection, player l claims that he can maintain at least his allocation in $(\mathbf{x}; \mathsf{X})$ by giving his new partners in Z at least what they had before in $(\mathbf{x}; \mathsf{X})$, and furthermore, if anybody in Z is also a member of Y, he gets at least what he would have gotten in Y.

An objection is *justified* if no counter-objection to it exists. Otherwise, it is *unjustified*. The bargaining set $\mathsf{M}_1^{(i)}$ is defined to be the set of all individually rational *p.c.*'s in which no player has a justified objection against any other member of the same coalition. The superscript of M indicates individual rationality and the subscript indicates that objections are made against one player at a time. Other bargaining sets have been proposed by Aumann and Maschler (1964), but for three-person quota games the results remain unchanged.

Aumann and Maschler showed that the bargaining set model provides a unique set of four *p.c.*'s in the three-person quota game, one for the case of no coalition (0, 0, 0; A, B, C) and one for each of the possible coalitions. The model does not provide a means of selecting which of the four will be chosen, or even a rationale for assuming equal likelihood.

C. *The Kernel*

The central concept of the kernel theory (Davis and Maschler, 1965) is that of *excess*. The excess of a coalition Y with respect to a *p.c.* $(\mathbf{x}; \mathbf{X})$ is given by

$$e(Y) = v(Y) - \sum_{i \in Y} x_i.$$

$e(Y)$ represents the total amount that the members of Y may gain or lose (depending on whether $e(Y)$ is greater or less than zero, respectively), if they desert the *c.s.* $\mathbf{X}$ and form the coalition Y. Clearly, $e(X_j) = 0$, $j = 1, 2, ..., m$, where $X_j \in \mathbf{X}$.

As before, let k and l be two players in a coalition X_j, and let $\tau_{k,l}$ be the set of all the coalitions which contain player k but not l, i.e.,

$$\tau_{k,l} = \{Y \mid k \in Y, l \notin Y\}.$$

The *maximum surplus* of k over l with respect to $(\mathbf{x}; \mathbf{X})$ is given by

$$s_{k,l} = \max_{Y \in \tau_{k,l}} e(Y).$$

This maximum surplus $s_{k,l}$ is the greatest amount that player k may gain (or the minimal amount he may lose, if $s_{k,l}$ is negative) by withdrawing from $\mathbf{X}$ and entering into coalition Y that does not include l, assuming that the other members of Y will be satisfied with receiving the same amount that they had had in $(\mathbf{x}; \mathbf{X})$. If the other members of Y are not content with their lot in $(\mathbf{x}; \mathbf{X})$, then $s_{k,l}$ may be viewed as the amount that player k has with which to bargain with the other players in Y in order to pursue more reward than his current allocation.

Player k is said to *outweigh* player l with respect to the *p.c.* $(\mathbf{x}; \mathbf{X})$ if $s_{k,l} > s_{l,k}$ and $x_l \neq 0$. A coalition is *balanced* if no player in it outweighs the other. The *kernel* of the game, denoted by $\mathbf{K}$, is the set of all individually rational *p.c.*'s having only balanced coalitions. It can be shown that $\mathbf{K} = \mathbf{M}_1^{(i)}$ for the three-person quota games considered in the present study.

Consider again our previous example. Suppose players A and B agree to split the value of their coalition 55:40. The respective *p.c.* is then

(55, 40, 0; AB, C).

Coalition AC can gain 35 points if AB is dissolved, but coalition BC can only gain 25 points, so player A outweighs player B in this configuration. By contrast, the maximum surplus coalitions AB or AC have over the $p.c.$

$$(0, 35, 30; A, BC)$$

is 60 points, so the coalition is balanced, and this $p.c.$ is in the kernel. The same holds for the $p.c.$'s $(60, 0, 30; AC, B)$ and $(60, 35, 0; AB, C)$.

D. *The Coalitions Game*

At the Thurstone Psychometric Laboratory of the University of North Carolina, we have created *Coalitions*, a set of programs written for a PDP-8 computer to play bargaining games within the characteristic function framework. Details of the program design and a discussion of its advantages in bargaining experiments may be found in Kahan and Helwig (1971).

The *Coalitions* game explicitly defines three stages in the bargaining process. The first, the *offer stage*, is that time when players explore the potentials of various coalitions. Different offers are made to different possible $c.s.$'s and some ideas towards where a reasonable solution to the game might lie are formed. In this stage, players gain awareness of their relative strengths and weaknesses, and some idea of the expectations of the other players.

The second stage, termed the *acceptance stage*, begins when a set of players indicate a general agreement on a division of the points. This agreement is not binding, but it does indicate that serious consideration of the $p.c.$ is in order. During the acceptance stage, the members of a tentative coalition may be interested in modifications of the agreement for that coalition as well as in what other $c.s.$'s might have in store for them. It is in this stage that various strategies and counter-strategies may become their most complex. For the student of bargaining behavior, it is the most interesting stage.

The third stage, or *ratification stage*, terminates the bargaining process. The members of a tentative coalition, having considered an offer and seeing it through acceptance, are now willing to make it a binding agreement. Satisfaction with the proposed coalition and its division of the points is therefore indicated and, by passing into ratification, each party receives its points and the bargaining (and game playing) process is terminated.

In the *Coalitions* game, communication among players takes place through the use of keywords, where a player specifies the type of message he wishes to send (the keyword), followed by the coalition to which it refers, and the allocation of the points to the members of that coalition. The basic keyword in the game is *offer*. A player may offer an allocation of the points to a coalition of which he is a member by stating *offer*, followed by an assignment of the points to both members of the coalition. Upon observing an offer made by another player that involves himself, a player may take a number of actions. First, he may indicate displeasure by sending the keyword *reject*, indicating which potential coalition he is rejecting. Second, he may indicate approval of a proposal by sending the keyword *accept*. Acceptance is understood to be approval to the most recent offer to the particular coalition indicated in the message. The game moves conceptually from the offer stage to the acceptance stage when one player has sent an *accept* of another's *offer*. Third, a player may make a counter-offer to the same coalition by sending an offer of his own. Only the most recent offer to any coalition structure is active at any one time. Fourth, a player may ignore the offer altogether and concentrate on other coalition structures. The offer remains in effect, but in limbo until acted upon.

When a *p.c.* has been in the acceptance stage long enough for at least two complete rounds of messages, it may enter the ratification stage, if a member of the coalition sends the keyword *ratify*. When *ratify* is sent, members of the coalition must immediately either agree to the ratification or veto it. In the former case, the game is ended with that particular *p.c.* in force; in the latter case, the game reverts back to the offer stage.

There are two further keywords used in the present experiment. If a player wishes to make no communication, he sends the keyword *pass*. If he wishes to leave the game with zero points, he sends the keyword *solo*, which immediately transforms the situation to a two-person game, excluding the *solo*-ed player.

One of the purposes of the present study is to test the bargaining set and kernel models by comparing prescribed and observed outcomes in three-person quota games. The various game-theoretic solutions concern themselves only with such outcomes, and not with how they came about. However, to limit the data analysis to this aspect, as previous n-person game experiments (*e.g.*, Maschler, 1965; Selten and Schuster, 1968) have done, would be to ignore the richness of the bargaining process leading to ratified outcomes and the possibility of determining how various coalitions are formed. Moreover, investigation of the bargaining process may lead to the discovery of effects of interpersonal bargaining, which may, in turn, result in modifications of mathematical models along psychological principles. In addition to these global goals, in the present experiment we shall also investigate the effects of characteristic functions that provide differing relationships among the three players, and also different means of communication which might have effects on the bargaining process and, consequently, on the outcomes as well.

II. METHOD

A. *Subjects*

Subjects were undergraduate males at the University of North Carolina recruited for this experiment via posted advertisements offering monetary rewards for participation in decision-making experiments. Subjects participated in groups of four. In no case were two players in one group close friends, although there were acquaintances who played together.

B. *Playing Procedure*

Each quartet of players was introduced to the experiment in one three-hour familiarization session. The first hour of this session was given to a written and verbal presentation of the rules of playing the game and how to operate a teletypewriter. The reader is referred to Kahan (1970) for this aspect of the procedure. Following this extensive introduction, subjects played practice games for the remaining two hours under the supervision of the experimenters, who often intervened to suggest lines of play in order to insure that subjects would be familiar with all of the technical details of the *Coalitions* game and the different types of situations that might arise in the course of a game. For this familiarization session subjects were paid $ 4.50 each.

In the succeeding three weeks, each quartet played three sessions of approximately three hours each, during which a total of twenty games per quartet were played. All games were three-person games, with one player of the quartet acting as a non-participating 'observer', so that each subject actually played in 15 games. This experimental procedure was employed not only to prevent subjects from knowing which of their compatriots were in which roles, but also to allow subjects time to reflect upon the task and to prevent them from knowing which of the other players were actually present in a particular game. This increased the validity of the assumption of the independence of different games within a quartet.

The incentive to maximize points was provided by paying each subject 5 cents for every point he gained in the 15 games that he played. This resulted in an average gain per subject of $25.00 per experiment, for a rate of somewhat over $2.75 per hour.

C. *Experimental Games*

Each quartet played five different three-person quota games, four times each. For each of the four iterations of a game, each member of a quartet played each position (including observer) once. The five games, shown in Table I, were chosen to represent different relationships among the quota values for the three players. In Game I there are two relatively poor players in contrast with a wealthy one. In Game II the situation is reversed,

TABLE I

Characteristic function and quota values for each game

Characteristic function[a]	Game				
	I	II	III	IV	V
$v(AB)$	95	115	95	106	118
$v(AC)$	90	90	88	86	84
$v(BC)$	65	85	81	66	50
Quotas					
ω_A	60	60	51	63	76
ω_B	35	55	44	43	42
ω_C	30	30	37	23	8

[a] $v(A) = v(B) = v(C) = v(ABC) = 0.$

with one poor player in relation to two wealthier ones. Games III, IV, and V represent low, moderate, and high differences, respectively, in quota values among all three players. The order of play of the five types of games was randomly selected. Within each game, the order of which player occupied which role was also randomly determined under the restriction that no person would be the observer for two consecutive plays (albeit of different games).

D. *Communication Variations*

Data are reported in the present paper for three separate experiments, each of which used the same five experimental games in the same order. Four quartets, or 16 players, participated in each experiment, for a total of 48 subjects. The differences among the three experiments are accounted for in terms of two communication variables. The first is whether or not secret offers were possible. If secret offers were possible, then one player might suggest to the other a split of their joint reward without the knowledge of the third, whose consent was not needed in any event. The option of allowing the third party to know about the offer is the sole prerogative of the originator of the message; the recipient does not know whether the offer was sent in secrecy or broadcast to all players. An offer, if made secretly, remained secret until it was tentatively accepted, at which time it became public. If secret offers were not possible, all messages were transmitted to all players. The second communication variable was whether or not subjects were allowed to transmit messages in a fixed order, or could speak at will upon request. In the *fixed order* variation, each of the three players communicated one message in turn, in an order fixed for each individual play and not dependent on the relationships of the quotas among the players. Within each experimental game, each player communicated first, second, and third one time. In the *ad lib* variation, subjects hit a special 'attention' key on the teletypewriter, and were allowed to communicate messages, one message per request, in the order in which they requested to speak.

The two communication variables were manipulated to yield the following three experimental treatments:

Exp. 1: no secret messages, speak in order.
Exp. 2: secret messages allowed, speak in order.
Exp. 3: secret messages allowed, speak at will.

III. RESULTS

The organization of the results section is based on successively deeper examinations of the data. We shall first look at the ratified outcomes, comparing them to the outcomes prescribed by the bargaining set and kernel models, and testing how they were affected by the communication rules governing the game, values of the characteristic functions, and learning. Second, we shall examine the negotiations in the initial phase of the experiment, particularly in the ultimately winning coalition, in order to trace the path of the ratified outcome. Thirdly, we shall analyze selected aspects of the bargaining process. A viewpoint and terminology based on the vocabulary of the bargaining set theory will be employed throughout the present section. Translation to other vocabularies should not be difficult for the industrious reader.

A. *Final Outcomes*

Tables II, III, and IV, present the outcomes (ratified payoff configurations) for each quartet, each game, and each iteration separately, for Experiments 1, 2, and 3, respectively. The last two rows of each table present the mean outcomes computed over the nonzero payoffs, and the corresponding quota values. An inspection of the three tables shows that within quartet differences in the coalition types that were actually formed and in the rewards obtained by their members are of the same type and magnitude as the respective differences among quartets. Also, post-experimental interrogation of the subjects indicated the success of our experimental procedure in preventing subjects from knowing which of the other players were actually present in each play. Hence, different games and repeated plays of the same game within quartet will be assumed independent in some though not all of our subsequent analyses.[1]

When considering the outcomes, the primary variables of interest are the incidence of formation of the various coalitions and the allocations of points for their members. Recall that the bargaining set and kernel models make no predictions as to which coalition should form. From this, it is possible to hypothesize that, given large numbers of independent repetitions of the same characteristic function game, each type of coalition (AB, AC, or BC) would occur as frequently as the others. In opposition to this 'equally likely' hypothesis, the social scientific tradition, through Caplow (1956, 1959) and Gamson (1961) predicts that the two weaker players, B and C, would combine against the stronger, A, resulting in a predominance of BC coalitions. Chertkoff (1967) predicts BC to AC coalitions in the ratio of 2:1, with no AB coalitions.

TABLE II
Outcomes of Experiment 1

Quartet	Game I			Game II			Game III			Game IV			Game V		
	A	B	C	A	B	C	A	B	C	A	B	C	A	B	C
1	0	47	18	58	57	0	55	40	0	81	25	0	68	50	0
	45	50	0	60	0	30	47	48	0	53	53	0	68	50	0
	60	0	30	65	50	0	0	49	32	60	0	26	59	59	0[a]
	60	35	0	65	50	0	48	47	0	53	53	0	59	59	0
2	55	40	0	0	50	35	50	45	0	0	50	16	68	50	0
	70	0	20	0	65	20	60	0	28	36	70	0	70	48	0
	0	0	0[a]	0	65	20	0	45	36	55	51	0	59	59	0[a]
	60	0	30	60	0	30	50	0	38	63	43	0	64	0	20
3	0	30	35	65	50	0	60	0	28	60	46	0	68	50	0
	60	0	30	70	0	20	0	40	41	56	50	0	79	0	5[a]
	50	45	0	80	0	10	44	0	44	66	40	0	62	56	0
	50	0	40[b]	50	65	0	48	47	0	54	52	0	60	58	0
4	0	37	28	70	45	0	0	40	41	55	0	31	68	50	0
	64	31	0	65	50	0	0	46	35	56	50	0	95	23	0
	70	0	20	75	0	15	55	40	0	75	31	0	80	38	0
	55	40	0	60	55	0	73	0	15	66	40	0	75	43	0
Mean	58.2	39.4	27.9	64.8	54.7	22.5	53.6	44.3	33.8	59.3	46.7	24.3	68.9	49.5	12.5
Quota	60	35	30	60	55	30	51	44	37	63	43	23	76	42	8

[a] Following *solo* by C.
[b] Following *solo* by B.

TABLE III
Outcomes of Experiment 2

Quartet	Game I			Game II			Game III			Game IV			Game V		
	A	B	C	A	B	C	A	B	C	A	B	C	A	B	C
1	55	40	0	60	55	0	0	45	36	66	40	0	75	43	0
	60	0	30	60	55	0	48	47	0	70	36	0	71	47	0
	65	0	25	0	55	30	0	40	41	53	53	0	74	44	0
	55	40	0	0	55	30	55	40	0	60	0	26	70	48	0
2	0	35	30	65	50	0	55	0	33	70	0	16	85	33	0
	60	35	0	0	60	25	0	50	31	0	50	16	73	45	0
	55	40	0	0	58	27	0	46	35	66	40	0	80	38	0
	53	42	0	65	50	0	0	45	36	70	36	0	59	59	0[a]
3	52	43	0	0	50	35	0	41	40	0	45	21	80	38	0
	48	47	0[a]	55	60	0	55	0	33	70	36	0	60	0	24
	55	0	35	65	0	25	0	45	36	66	0	20	59	59	0[a]
	55	40	0	0	55	30	58	0	30	60	46	0	72	46	0
4	58	37	0	0	50	35	50	0	38	66	40	0	59	59	0
	55	40	0	0	60	25	48	0	40	63	43	0	85	33	0
	65	0	25	75	40	0	58	0	30	0	35	31	78	40	0
	62	33	0	63	52	0	60	35	0	65	41	0	78	40	0
Mean	56.9	39.3	29.0	63.5	53.7	29.1	54.1	43.4	35.3	65.0	41.6	21.7	72.4	44.8	24.0
Quota	60	35	30	60	55	30	51	44	37	63	43	23	76	42	8

[a] Following *solo* by C.

TABLE IV
Outcomes of Experiment 3

Quartet	Game I			Game II			Game III			Game IV			Game V		
	A	B	C	A	B	C	A	B	C	A	B	C	A	B	C
1	48	47	0	0	60	25	55	0	33	80	0	0	73	45	0
	55	40	0	0	45	40	55	0	33	0	51	15	70	48	0
	60	35	0	65	50	0	58	0	30	58	48	0	60	0	24
	70	25	0	55	0	35	55	40	0	76	0	10	59	59	0[a]
2	55	40	0	60	55	0	47	48	0	53	53	0	88	30	0
	0	31	34	0	60	25	50	45	0	58	48	0	70	48	0
	53	42	0	65	50	0	50	45	0	61	45	0	68	50	0
	55	40	0	67	48	0	0	38	43	60	46	0	65	53	0
3	60	0	30	60	55	0	55	40	0	60	46	0	70	48	0
	55	40	0	60	55	0	0	36	45	66	40	0	75	43	0
	65	0	25	65	50	0	58	0	30	63	43	0	80	38	0
	66	0	24	60	55	0	50	45	0	0	50	16	68	50	0
4	63	0	27	59	56	0	50	45	0	56	50	0	64	0	20
	53	0	37	57	58	0	55	40	0	53	53	0	74	0	10
	55	40	0	0	70	15	0	46	35	70	36	0	100	18	0
	45	0	45	55	0	35	58	0	30	53	53	0	92	26	0
Mean	57.2	38.0	31.7	60.7	54.8	29.2	53.5	42.5	34.9	61.9	47.3	11.8	73.5	42.8	18.0
Quota	60	35	30	60	55	30	51	44	37	63	43	23	76	42	8

[a] Following *solo* by C.

The frequencies of coalition types were tabulated over the five games and three experiments. As no differences due to Experiment were revealed, the frequencies were collapsed over communication differences. As Table V shows, there is a decided tendency for the stronger players to team up and share the largest reward offered. The preeminence of the formation of the AB coalitation does not occur equivalently over games, however, but differs significantly from game to game ($\chi_4^2 = 32.683$, $p < 0.001$). Because of the possibility that repeated plays of the same game might bias results, the first play of each game was also examined in similar fashion, with results essentially identical to those repeated.

TABLE V
Frequency of coalition structures by game
Coalition Structure

Game	AB, C	AC, B	A, BC	A, B, C
I	27	15	5	1
II	25	8	15	0
III	17	16	15	0
IV	35	7	6	0
V	42	6	0	0
Over game	146	52	41	1

The deviations of the mean outcome from the predicted quota values are easily computable from the last two rows of Tables II, III, and IV. Overall, these deviations are small, and as later analyses will show, not significantly different from zero. To make results comparable over games and coalition types, define a deviation score for each $p.c.$, whether at the offer, acceptance, or ratification stage, as the number of points proposed to the stronger player in the two-player coalition (i.e., A in AB, A in AC, and B in BC) minus his quota. Thus, a positive number indicates that the stronger player in a coalition received more than his predicted quota, whereas a negative number indicates that the allocation gained the stronger player less than the prescribed quota value. The majority of all instances of negative deviations indicate $p.c.$'s more egalitarian than the quota predictions; only a few of these instances indicated a disposition to the absolute advantage of the weaker player. Table VI shows the mean deviation scores for the outcomes for each game type and each coalition type separately.

The effects of three factors on the deviation scores were studied: Experiment (with 3 levels), Game (with 5 levels), and Iteration (with 4 levels). The four repetitions of each game constituted the latter factor. Because both the Game and Iteration factors were repeated over quarterts, the mean deviation scores were transformed to orthogonal contrasts of these factors and analyzed in multivariate fashion.[2] No main effects, in-

TABLE VI

Mean deviation from quota of first offers to the winning coalition, first accepted offers by that coalition, and outcomes by game and coalition type

Coalition	Game					Over games
	I	II	III	IV	V	
	First offers to winning coalition					
AB	−5.38	2.16	−0.12	−6.17	−8.62	−4.59
AC	−4.19	4.38	5.56	−4.00	−11.50	−0.74
BC	−1.00	−0.67	1.87	−2.00	[a]	0.02
Over coalition	−4.51	1.65	2.40	−5.33	−8.98	−2.95
	First accepted offers by winning coalition					
AB	−4.12	2.56	1.29	−2.80	−3.81	−1.92
AC	−2.13	5.00	8.06	1.71	−11.50	1.47
BC	−1.00	2.07	−0.87	2.33	[a]	0.66
Over coalition	−3.11	2.81	2.88	−1.50	−4.77	−0.74
	Outcomes					
AB	−4.33	2.36	0.65	−2.03	−3.74	−1.88
AC	0.47	5.00	4.94	3.71	−9.17	1.87
BC	1.00	2.20	−0.53	3.83	[a]	1.29
Over coalition	−2.23	2.75	1.71	0.46	−4.42	−0.52

[a] No cases observed.

cluding the grand mean effect (overall difference from zero), or interactions were significant ($\alpha = 0.05$), except the main effect for Game ($F_{4,6} = 14.54, p < 0.003$). Consequently, in view of this result and the results for frequency of the c.s.'s presented in Table V, the data are presented collapsed over Experiment and Iteration.

Inspection of Table V and the lower part of Table VI suggests that the five different games may be grouped into two classes, with Games I, IV, and V in one group, and Games II and III in the other. As Table V shows, the number of *BC* coalitions formed in Games I, IV, and V was 5, 6, and 0, respectively, whereas the number of *BC* coalitions for Games II and III was 15 for each. The lower part of Table VI shows that the stronger player in a coalition received less than his quota in Games I, IV, and V, but obtained more than his quota in Games II and III. The latter overall pattern is shown in the AB coalition case, and not in *AC* and *BC*, but recall that more than 60% of the final coalitions fell in the former category.

Further evidence concerning the difference between the two classes of games is provided by two measures of the intensity of bargaining activity, namely, the number of messages sent and the time spent per play. These two measures should, of course, be strongly correlated with each other. Multivariate analyses of variance identical to those performed on the

TABLE VII

Means and standard deviations of number of messages sent and time spent per play by game and iteration

	Number of messages		Time of play (in minutes)	
	Mean	S.D.	Mean	S.D.
Game				
I	20.96	13.50	18.00	14.45
II	18.92	9.79	16.15	12.76
III	19.10	7.33	16.00	8.59
IV	21.63	11.18	17.75	11.27
V	25.19	13.82	19.73	13.13
Iteration				
1	25.78	13.26	24.72	15.64
2	21.38	12.93	16.90	12.57
3	18.17	7.01	14.17	6.03
4	19.25	10.48	14.32	9.34

deviation scores were performed on these two measures. Significant ($\alpha = 0.05$) effects were found for Game and Iteration only. (For number of messages sent per play, $F_{4,6} = 18.80$, $p < 0.002$ for Game; and $F_{3,7} = 10.75$, $p < 0.005$ for Iteration. For time spent per play, $F_{4,6} = 6.40$, $p < 0.023$ for Game; and $F_{3,7} = 12.61$, $p < 0.003$ for Iteration.) No other significant main effects or interactions were found.

The mean and standard deviation of number of messages sent and time spent (in minutes) per play are shown in Table VII for Game and Iteration. As expected, the two measures are highly correlated. The differences in bargaining intensity for Game follow the results for frequency of coalition type and allocation of points in that Games II and III are distinctly different from the other three games. The Iteration effect, shown in Table VII, is clearly a learning effect. The first time a game is played, much more time is spent and more messages are sent as the players learn the features of that particular game. By the third and fourth iterations, there is little change in mean bargaining intensity. This does not mean, however, that the players returned to the same *p.c.* on repeated iterations of a game; even cursory inspections of Tables II, III, and IV indicate the opposite.

According to the rules of the game, players had the opportunity of withdrawing from play with zero points by playing *solo*. If a player did use *solo*, the situation was transformed to a two-person cooperative game. Out of 240 plays *solo* occurred a total of nine times, with three quartets employing it once, and three quartets employing it twice. Three *solo*'s occurred in Game I, and the remainder in Game V. These outcomes are footnoted in Tables II, III, and IV. In all cases but one (Game I), it was player C who chose *solo*. The outcome following a *solo* was generally an equal division of the points between the two remaining players. If these nine cases are deleted from consideration, the means for Games I and V become less egalitarian, particularly in Experiment 1, but the overall patterns and significances remain unchanged. The recorded incidence of *solo* lends additional support to the previous interpretation of the final outcomes in terms of the differential saliency of coalition types in different games.

B. *The Initial Phase of Negotiations*

The deviation from quota score has served as the main dependent variable for assessing the effects of Experiment, Game, and Iteration on the final disbursement of rewards. Deviation from quota scores may also be computed for the first offer made to the coalition that ultimately became established (the *winning* coalition), and for the offer that was first accepted by both members of that coalition. (The two offers may or may not coincide.) Some insight into how the final disbursement of rewards was determined may be gained from comparing the deviation scores for the first offer made to the winning coalition, the first offer accepted by that coalition, and the offer ultimately ratified by that coalition.

A multivariate analysis of variance, identical to the one performed on the deviation scores for the outcomes, was conducted for the deviation scores for the first offer made to the winning coalition. There was a significant grand mean effect ($F_{1,9} = 11.80$, $p < 0.007$), showing that the bargaining set and kernel models, while predicting the outcomes, do not account for the first offers made to the winning coalition. The other significant effect was due to Game ($F_{4,6} = 8.61$, $p < 0.012$). The multivariate Iteration effect was nearly significant ($F_{3,7} = 3.92$, $p < 0.062$), and the univariate linear trend across iterations was highly significant in this case ($F_{1,9} = 10.01$, $p < 0.011$). No other significant main effects or interactions were observed.

A multivariate analysis of variance performed on the deviation scores for the first offer accepted by the winning coalition yielded a significant Game effect ($F_{4,6} = 8.08$, $p < 0.014$). There was also a significant Experiment × Game × Iteration effect ($F_{12,8} = 4.90$, $p < 0.016$), which has no immediate explanation and is probably due to chance. No other significant main effects or interactions were observed. In particular, the grand mean effect, which was significant for the first offer made to the winning coalition, was not significant for the first accepted offer by that coalition ($F_{1,9} = 1.35$, $p < 0.307$).

The upper two parts of Table VI present the mean deviations from quota by Game and type of coalition for the first offer to the winning coalition and for the first accepted offer by that coalition. With minor exceptions, the patterns of deviation from quota between games parallels exactly that of the outcomes, which are shown in the lower part of Table VI.

A comparison of the grand mean effects for the deviation scores ($F_{1,9} = 11.80$, 1.35, and 0.61, for the first offer made to the winning coalition, the first accepted offer by that coalition, and the ultimately ratified offer by that coalition, respectively) shows that the mean overall allocation of rewards approached the quota as play progressed. Examination of Table VI shows that this effect is one of moving away from egalitarian solutions toward the predicted quota values. This tendency is partially attributable to the differential Iteration effect, which is present for first offers made to the winning coalitions, but not for first or final acceptances by those coalitions. It appears likely that players enter the game with some sharing or equity norm but that this norm is rapidly abandoned as the power relationships inherent in the game become realized.

The right-hand column of Table VI shows that the progressively less egalitarian disbursement of rewards holds for each of the three coalition types. In the AB winning coalitions the mean difference for player A from his quota was -4.59 points for the first offer, -1.92 points in the first accepted offer, and -1.88 points in the ratified coalition. In the AC winning coalitions the respective mean differences from the quota for player A were -0.74, 1.47, and 1.87, showing the same rate of increase as in coalition AB. The rate of increase for player B in the BC winning coalitions was more moderate; his respective mean differences from quota, all of which are positive, were 0.02, 0.66, and 1.29.

A multivariate analysis of variance was also performed on the deviation scores for the first offer made in each play, regardless of to which coalition, winning or losing, it was addressed. The mean deviation scores were significantly different from zero ($F_{1,9} = 311.53$, $p < 0.001$), with the grand mean effect being considerably larger than in the case of the first offer made to the winning coalition. The other significant effects were due to Game, Iteration, and their interaction. In particular, no significant main or interaction effects due to Experiment were found, showing that the two communication variables manipulated in the present study had no effect not only on the outcomes but also not on the initial offers.

C. *The Bargaining Process*

The bargaining set and kernel models concern only the outcomes, and do not concern themselves with the characterization of the bargaining process. From a psychological viewpoint the latter is of much more interest. The two models may be represented as sets of solutions, each based on different assumptions, of systems of linear inequalities involving the rewards allocated to members of the ratified coalition as unknowns. It is obvious that our human game players do not set up in their minds systems of linear inequalities and solve them in order to determine how to disburse the reward jointly gained by members of a particular coalition. Instead, they employ threats, counter-threats, promises, bluffs, and other negotiation tactics in proceeding to the ratification of a coalition and the disbursement of its reward.

The analysis of the negotiation moves appearing in the game protocols to which we now turn, is divided into two parts. The first part, a static analysis, examines several useful indices of bargaining which may be affected by the Experiment, Game, and Iteration factors. These indices include the number of offers made by each player, the number of accepted offers (tentative coalitions) per play, and the message number of (how many messages were sent before) the formation of the first tentative coalition. The latter part is a dynamic analysis, based upon a model of the bargaining process, which examines the offers sent during a play in light of the offers that precede or follow them.

1. *Static analysis*. Multivariate analyses of variance, similar to the ones conducted above, were performed on the message number of the first tentative coalition and on the number of tentative coalitions formed during a play. These analyses showed no significant effects involving Experiment, Iteration, or Game for either of the two dependent variables. Again, the type of coalition was not considered as a factor in either analysis. When the message number of the first tentative coalition and the total number of tentative coalitions formed per play were examined according to which coalition was first formed, a different picture emerged. Although the message number of the first tentative coalition did not vary with the type of coalition, it was shown that if coalition AB was the first to form, there were fewer tentative coalitions following it than if either of the other two coalitions formed first. The mean and standard deviation of the number of tentative coalitions per play were 2.22 and 2.03 when coalition AB was

TABLE VIII

Percentages of offers by game, originator of offer, and target of offer

From	To	I	II	III	IV	V	Over game
A	B	54	61	61	70	71	64
A	C	46	39	39	30	29	36
B	A	66	61	53	71	78	66
B	C	34	39	47	29	22	34
C	A	69	49	56	59	66	60
C	B	31	51	44	41	34	40
[a]	Higher	64	56	56	66	71	63
[b]	Lower	36	44	44	34	29	37

[a] From C to B, B to A, and C to A.
[b] From A to C, B to C, and A to B.

first formed, 3.32 and 2.52 when coalition AC was first formed, and 3.15 and 2.64 when coalition BC was first formed, the difference being significant by analysis of variance ($F_{2,237} = 6.25, p < 0.002$).

The third index of bargaining to be analyzed was the number of offers sent by each player per play. Table VIII shows the percentages of offers sent by each player to each of the other two players broken down by Game, originator of offers, and target of offer. Table VIII shows that players were more likely to address the stronger of the other two players than the weaker player. This effect, however, seems to be related to the particular game played. It is less prominent in the two all-coalition-active games (II and III) than in the other three games, following the same pattern of differences between the two classes of games detected above. The individual players conformed to this communication preference pattern except for player A in Game I, who showed indifference in making an offer to his two almost equally weak potential partners.

2. *Dynamic analysis.* The use of a dynamic analysis of the bargaining process, which examines messages sent during a play in light of messages that precede or follow them during the same play, requires the specification of a model to classify the various messages and describe the progress of the negotiations. In choosing among various alternatives for modelling the bargaining process, we were motivated by the success of the bargaining set and kernel models in accounting for the outcomes of the present experiment. The constraint on any proposed model that the bargaining process converge to the bargaining set or kernel solutions considerably narrowed the range of alternatives. As a first alternative, extensions of either of the two static models were considered, through providing testable dynamic interpretations to their assumptions. The bargaining set model was chosen over the kernel model because its terminology is more immediately representable in observable responses, and, from a strictly practical viewpoint, because it accounted for the outcomes in another study (Horowitz and Rapoport, 1974), whereas the kernel model did not.

Our proposed model follows the path of a particular *p.c.* from its initial consideration for adoption through its being tested against other alternatives to its final acceptance or rejection. The proposed structure of the bargaining process, the same one considered by Horowitz and Rapoport (1974), is as follows: After preliminary negotiations, the bargaining process enters the stage when a tentative coalition X is formed. The tentative coalition may either result from the initial negotiations among the three players or may follow the dissolution of a previous tentative coalition. Players are assumed to search for justified objections against X. If none exist, the coalition is eventually ratified. If at least one justified objection exists, then one of them is selected by a player, and is expressed through an offer (*p.c.*). We denote such an expressed objection, whether justified or unjustified, as Y_s.

The rules of the *Coalitions* game dictate that these objections be made serially, $s=1, 2, ...$, and that the payoff vectors be integers rather than real numbers. The identity of the player expressing the objection is not considered here. Thus, for example, if a tentative coalition $X=AB$ is formed, an objection $Y_s=AC$ by A against B may be made by A through an offer to C, or by C through an offer to A. This requirement is compatible with the definition of objection (Aumann and Maschler, 1964).

If the objection Y_s is justified, the tentative coalition X is immediately dissolved and the objection is accepted, thus resulting in a new tentative coalition, and the beginning of the cycle over again. If, however, the objection is unjustified, the tentative coalition X is retained, and a new objection against X, denoted by Y_{s+1}, is made. A player may express both unjustified and justified objections. If at least one of the players possessing a justified objection against X eventually expresses that objection, and we assume that he will, it can be shown that, for the class of games considered in the present study, this bargaining model results in convergence to a *p.c.* in the bargaining set.

Not all offers fit within the definition of an objection, either justified or not. The first offer and all offers before the first acceptance are not objections. Also, if fewer points are offered to a player in a tentative coalition than he would receive in that coalition, the offer is not an objection. The bargaining process model posits that these latter non-objections should not occur; in the present data they occur only rarely.

To exemplify the bargaining process leading to $M_1^{(i)}$, consider Game I. Supposing that after coalition AB is formed with a *p.c.* (55, 40, 0; AB, C), C offers 58 points to A. Since this is a justified objection by A against B, coalition AB is dissolved and the *p.c.* (58, 0, 32; AC, B) is formed after A accepts C's offer. Next, for example, B may propose 65 points to A, i.e., (65, 30, 0; AB, C), which is an objection by A against C. However, it is not justified since C may counter-object with an offer such as $(0, 33, 32;$ $A, BC)$. Hence the game proceeds with coalition AC remaining intact. Finally, for example, supposing A proposes the $p.c.$ $(60, 35, 0; AB, C)$. Since this is both a justified and stable objection by A against C, C cannot counter-object and no justified objections may be made against it. Thus, it becomes ratified.

The bargaining process model described above fails its most critical test; all games did not end in the predicted quota values. While reserving discussion of this point for later, useful information may be still obtained from examining the bargaining process to discover the sources of discrepancy from the model, which may suggest how to modify it. Within this context, then, we shall test the bargaining process model by examining several statistics relative to objections.

An enumeration of all offers classifiable as objections was made, and each objection was examined to see whether it dissolved the tentative coalition against which it was addressed (*accepted*) or was unsuccessful in that attempt (*unaccepted*). An objection is said to have been unaccepted if either the tentative coalition to which it was addressed was later ratified, or if the tentative coalition was disrupted by a subsequent objection, involving either the same or a different coalition structure.

TABLE IX

Percentages of objections by game, type of objection, and nature of outcome

Objection			Objection is later	
Game	Type	n	Accepted	Unaccepted
I	J	65	40	60
	UJ	125	22	78
II	J	57	32	68
	UJ	123	15	85
III	J	58	48	52
	UJ	152	20	80
IV	J	87	33	67
	UJ	124	28	72
V	J	119	27	73
	UJ	72	26	74
Over game	J	386	35	65
	UJ	596	22	78

Table IX presents the effect of the two types of objections on tentative coalitions for each game separately and over games. Examination of the table shows that justified (J) objections were more effective in dissolving tentative coalitions than unjustified (UJ) ones; 35% of all justified objections disrupted the tentative coalition, as compared to 22% for the unjustified ones. This difference between justified and unjustified objections appears in Games I, II, III, but not in IV and V. This difference between games is particularly pronounced when the fate of the justified objections is examined. In Games I, IV, and V, where *AB* was the dominant coalition, the disrupting justified objections were more likely to be in turn disrupted than not (the corresponding percentages are 31 and 9 for Game I, 23 and 10 for Game IV, and 18 and 9 for Game V). However, in Games II and III a justified objection that was tentatively accepted was as likely as not to be ratified.

Table X shows the same percentages of objections as Table IX, broken down by iterations instead of by games. The table shows a marked learning effect for the differentiation of justified from unjustified objections. There is no difference in the effect of both types of objections in the first iteration. However, starting with iteration 2, the percentage of justified objections dissolving the tentative coalitions increases slowly (the percentages are 29, 36, 37, and 41 for iterations 1, 2, 3, and 4, respectively), whereas the percentage of unjustified objections disrupting the tentative coalitions slightly decreases.

The justified and unjustified objections were further broken down by the coalition they were addressed to, by whether the objection was originated by a member of the tentative coalition (an *inside* player) or by the player excluded from it (the *outside* player), and by whether the objection was addressed to the more or less powerful members of the tentative coalition. The number of objections lodged against coalition *AB* was greater than for the other two coalition types, because more *AB* coalitions were formed than the other types. Because there was no interaction between the coalition type and the source and target of the objection, results were collapsed over those variables.

Table XI shows the incidence of justified and unjustified objections by Game, source of objection (inside or outside), and target of objection. It can be seen immediately that the outside player made four to five times more objections per game, either justified or not, than an inside player, who made offers to the other member in the tentative coalition in order to better his own position. For the justified objections, but not for the unjustified ones, the stronger player was more likely to be addressed in the objection, regardless of its source. Further inspection of Table XI shows that the latter effect was most pronounced for Games I, IV, and V, where player A had a very clear edge over the other players, but not in Games II and III. In Game III, in which the differences among quota values were the smallest, it was the weaker rather than the stronger player who was more likely to be addressed in the objection, regardless of its source. Unlike the justified objections, the unjustified objections were addressed

TABLE X

Percentages of objections by iteration, type of objection, and fate of objection

Objection			Objection is later	
Iteration	Type	n	Accepted	Unaccepted
1	J	156	29	71
	UJ	181	26	74
2	J	77	36	64
	UJ	129	24	76
3	J	70	37	63
	UJ	140	16	84
4	J	81	41	59
		148	20	80

about evenly to the stronger and weaker players, regardless of the source of the objection.

A final examination looks at the success or failure of an objection to disrupt the coalition it is addressed to, based on the actual offer that the objection constitutes. Objections were classified into one of five categories, depending on whether the more powerful player was offered (i) less than 10 points off his quota, (ii) between 5 and 10 points smaller than his quota, (iii) within 4 points of the quota, (iv) between 5 and 10 points greater than his quota, or (v) more than 10 points greater than his quota. Table XII presents a frequency of objections separated by this classification of deviation from quota scores, by Game, and by whether the originator of the objection was an inside or outside player.

Immediate examination of the middle part of Table XII shows that, except for Game V, more objections were in categories (iv) and (v) than in (i) and (ii). In Game V, for player C, who was concerned with most of the objections, there was a maximum limit of 8 points he could give to more the powerful player, which explains in part the difference between Game V and the remaining four games for outside objections. Because of the generally low frequency of inside objections per game, it is not possible to clearly determine between game patterns on total number of objections.

Turning to the percentage of objections accepted, the immediately obvious finding is that inside objections were accepted more than twice as often as outside objections. Of more interest is the finding that (again with the exception of Game V) the percentages of accepted objections constitute a single-peaked, unsymmetric function, with a mode in category (ii) or (iii). This finding holds for outside objections but not for inside objections, showing that, for the former class, objections in the -5 to -10 class (for Game IV) or in the -4 to $+4$ class (for Games I, II, and III) were more likely to be accepted than objections farther away from the quota, even if these farther away proposals were more in favor of the inside member of the objecting coalition.

TABLE XI

Incidence of justified and unjustified objections by game, source of objection, and target of objection

Game	From inside player against		From outside player against	
	Stronger player	Weaker player	Stronger player	Weaker player
	Justified objections			
I	11	5	38	11
II	8	8	28	13
III	3	11	18	26
IV	15	4	51	17
V	18	4	80	17
Over game	55	32	215	84
	Unjustified objections			
I	10	11	63	41
II	10	8	49	56
III	15	10	73	54
IV	8	7	58	51
V	11	9	21	31
Over game	54	45	264	233

TABLE XII

Analysis of fate of objections by game, origin of objection, and deviation from quota scores

Game	Score	Inside objections			Outside objections			All objections		
		Acc.	Total	% Acc.	Acc.	Total	% Acc.	Acc.	Total	% Acc.
I	< −10	1	3	0.33	2	10	0.20	3	13	0.23
	−5 to −10	7	10	0.70	11	37	0.30	18	47	0.38
	4 to −4	9	16	0.56	13	42	0.31	22	58	0.38
	5 to 10	3	4	0.75	3	38	0.08	6	42	0.14
	> 10	2	4	0.50	2	26	0.08	4	30	0.13
II	< −10	0	5	0.00	1	23	0.04	1	28	0.04
	−5 to −10	3	6	0.50	4	23	0.17	7	29	0.24
	4 to −4	2	4	0.50	6	29	0.21	8	33	0.24
	5 to 10	7	15	0.47	8	45	0.18	15	60	0.25
	> 10	2	4	0.50	3	26	0.12	5	30	0.17
III	< −10	2	5	0.40	0	26	0.00	2	31	0.06
	−5 to −10	3	3	1.00	3	20	0.15	6	23	0.26
	4 to −4	10	18	0.56	21	54	0.39	31	72	0.43
	5 to 10	4	7	0.57	8	26	0.31	12	33	0.36
	> 10	1	6	0.17	6	45	0.13	7	51	0.14
IV	< −10	0	3	0.00	5	23	0.22	5	26	0.19
	−5 to −10	3	5	0.60	13	31	0.42	16	36	0.44
	4 to −4	11	15	0.73	14	53	0.26	25	68	0.37
	5 to 10	2	5	0.40	8	35	0.23	10	40	0.25
	> 10	1	6	0.17	7	35	0.20	8	41	0.20
V	< −10	3	6	0.50	15	39	0.38	18	45	0.40
	−5 to −10	2	4	0.50	6	24	0.25	8	28	0.29
	4 to −4	7	12	0.58	10	65	0.15	17	77	0.22
	5 to 10	1	2	0.50	1	15	0.07	2	17	0.12
	> 10	3	18	0.17	3	6	0.50	6	24	0.25
Over game	< −10	6	22	0.27	23	121	0.19	29	143	0.20
	−5 to −10	18	28	0.64	37	135	0.27	55	163	0.34
	4 to −4	39	65	0.60	64	243	0.26	103	308	0.33
	5 to 10	17	33	0.52	28	159	0.18	45	192	0.23
	> 10	9	38	0.24	21	138	0.15	30	176	0.17

The latter analysis suggests that there are two opposing forces which affect the disruption of a tentative coalition. The first and obvious one is the magnitude of temptation to dissolve the coalition, which is a function of the magnitude of increment to the inside player of the objecting coalition. It remains to be seen whether the percentage of accepted objections is a monotonically increasing function of the magnitude of this increment or not. Additional analyses, not reported here, suggest the latter, with the function slowly increasing and then more quickly decreasing as the magnitude of increment to the objecting player is perceived by him as suspiciously large. The second force is directly related to the quota values. Table XII shows that (with the exception of Game V) the viability of an objection decreases as the absolute value of the deviation from quota score increases.

IV. DISCUSSION

Two separate questions arise in connection with cooperative n-person games with more than two players (Rapoport, 1971). The first question, which has been considered to be the more prominent by social scientists interested in coalition formation and bargaining, is: How will the n players organize themselves into coalitions, providing that the rules of the game allow (or require) coalitions? The second question, which has attracted the attention of mathematicians rather than social scientists, is. What will be the final distribution of payoffs? To this brace of questions we add a third one, seldom presented in the literature, which concerns itself with the bargaining process rather than its end result: How does bargaining progress from the initial offer to the final agreement? Within any of these three basic questions, there might arise questions of the effects of different characteristic functions, different modes of communication, and different experimental paradigms.

A. *Incidence of Different Coalitions*

With regard to the first question, two findings are pertinent. The first is the frequency of coalition structures as presented in Table V. The second is the highly significant difference between the number of tentative coalitions in plays in which coalition AB was the first to form and the number of tentative coalitions in plays in which either coalition AC or BC was first formed. Overall, coalition AB was the ultimate result of the bargaining in more than 60% of all plays, a result that stands in sharp contrast to the hypothesis of equal frequencies and in even sharper contrast to the hypothesis of predominance of *BC* coalitions (with perhaps some *AC* coalitions) that results from the social scientific theories currently in vogue. Not only was coalition *AB* the most frequently ratified, but it was also the most stable during bargaining. That is to say, the *AB* coalition, in addition to being the most likely to form, was the most likely to survive onslaughts of objections from inside or outside players.

It might be argued that the pre-eminence of the *AB* coalition results from the highest reward it provides to the players, and that the game is perceived by the subjects as a four-person, zero-sum game. A tacit coalition of subjects against experimenter would result in forming the *AB* coalition no matter who was player *A* and who was player *B*. However, we are disinclined to accept this rationale for two reasons. The first is the observed difference in the rate of *AB* coalitions among different games. This difference was not based on the difference of the *AB* coalition in potential gain versus the other two coalitions, but rather the differential quota of the *A* player (whose position is reflected in both the *AB* and *AC* coalitions rather than just the former). If the extra value of the *AB* coalition is the salient feature in its formation, than the frequency of its formation would be greater in Game II than in Game I. Our results in Table V indicate just the opposite. The second reason is that, once the three players tacitly form a coalition against the experimenter, the payoff among the three assumes some characteristics of an n-person Prisoner's Dilemma in that, by cooperating all players can do relatively well, but, by defecting, one player can, while remaining anonymous, do even better. Evidence from research on n-person Prisoner's Dilemma games (Marwell and Schmitt, 1972; Kahan, 1973) indicates that anonymous n-person Prisoner's Dilemma games are even more competitive in tone than their generally competitive two-person counterpart games. Thus, it would appear unlikely that subjects in this situation of not being directly responsible for their own actions would be so universally cooperative. Once one of the four players in a quartet attempts to exploit the other, then the agreement, being implicit to begin with, would quickly dissolve.

Even before considering the role of the payoff configurations as determining which coalition forms, we can fairly readily explain the differences between the present results and previous data supportive of Caplovian coalition theory and its offshoots. In previous experiments, the characteristic function was zero-sum. Any coalition won a constant value, and, in the absence of a coalition (and in contrast to the political convention model often claimed for such games), player A gained the payoff for himself. In addition, the formation of the coalitions and the dividing of the rewards were not unified, with the former occurring first. In such a situation it appears natural for the two weaker players to stop the stronger from winning all on his own before they consider how to divide up the rewards themselves. Once they have a coalition, player A is not really given the opportunity to intrude into their bargaining processes to make objections, so the BC coalition is ratified. As the present experiment is nonzero-sum, and as it does give the outside player the opportunity to take part in the negotiations, it is natural that the results should be different.

Despite differences in experimental procedure, Riker's (1967) experiment is sufficiently close in form to the present one that comparisons of data may be meaningfully made. Riker employed a single quota game with the characteristic function $v(A) = v(B) = v(C) = v(ABC) = 0$, $v(AB) = \$6.00$, $v(AC) = \$5.00$, and $v(BC) = \$4.00$. Since each point in our experiment was known to the subjects to be worth 5 cents, Riker's characteristic function may be rewritten for purposes of comparison, as equivalent to $v(AB) = 120$, $v(AC) = 100$, and $v(BC) = 80$ in conformity with the functions presented in Table I. Thus, Riker's quota values become $\omega_A = 70$, $\omega_B = 50$, and $\omega_C = 30$. This is an equal quota difference game, in which $\omega_A - \omega_B = \omega_B - \omega_C$, directly comparable to our three equal quota difference games, III, IV, and V. Equal quota difference games may be compared to one another in terms of an index of interplayer distance $\Delta = (\omega_A - \omega_B)/\omega_A$. In Riker's experiment $\Delta = 0.285$, whereas in our experiment Δ equals 0.137, 0.317, and 0.447 for Games III, IV, and V, respectively. Thus, Riker's quota game falls between Games III and IV in terms of the index Δ. The percentage of games played by undergraduate students (Groups II and III) in Riker's experiment in which coalition AB was formed was 0.526 (30 of the 57 games in which one of three coalition types, AB, AC, or BC, was formed). From our Table V, the respective percentages for Games III, IV, and V are 0.354, 0.729, and 0.875, with the percentage of AB coalitions for Riker's experiment falling in the appropriate interval. The same result holds for coalitions AC and BC.

The percentages of games in which coalitions AC and BC were formed in Riker's experiment were 0.316 and 0.158, respectively. The corresponding percentages for Game III are 0.333 and 0.313, for Game IV 0.146 and 0.125, and for Game V 0.125 and 0, with the percentage for Riker's experiment falling in appropriately.

B. *Effects of Independent Variables*

The comparison between Riker's experiment and the present one may be extended to the payoff configurations as well as the coalition structures of the outcomes. In 22 instances out of 93, the outcomes of Riker's experiment fell exactly in the bargaining set and kernel. Since his characteristic function involved only numbers divisible by 5, the only games of the present study directly comparable to Riker's game when testing the point predictions of both models are Games I and II. In these two games, as Tables II through IV show, the final outcomes were included in $M_1^{(i)} = K$ in 19 out of 96 plays. Moreover, "The average amount won, when player i was a winner, is almost exactly the quota for i" (Riker, 1967, p. 648). The same conclusion has been reached in the present study, as evidenced by the non-significant grand mean effect for the outcomes when measured as deviations from quota scores.

A comparison of several multivariate F values with their associated significance levels reveals an increasing tendency of the $p.c.$'s to approach $M_1^{(i)} = K$ as the game progresses. There are four grand mean F values, resulting from the multivariate analyses of the deviation from quota scores, which are pertinent for this comparison. They are $F = 311.53$ ($p < 0.001$) for the first offer made in a play, $F = 11.80$ ($p < 0.007$) for the first offer made to the winning coalition, $F = 1.35$ ($p < 0.307$) for the first tentative coalition formed by the winning players, and $F = 0.61$ ($p < 0.455$) for the outcomes. The corresponding mean deviation from quota scores are -4.31, -2.95, -0.74, and -0.52, respectively. Clearly, these four F values are not mutually independent, since included in the first offers were those made to the winning coalition, many of the first offers made to the winning coalition were also the first to be accepted, and many of the first tentative coalitions were later ratified. A comparison of the associated means shows, however, that as a play unfolded the $p.c.$'s became progressively less egalitarian, approacning the quota prediction.

Related to the within play learning discussed immediately above is the between play learning reflected in the iteration effect. Again, there are four sets of F values with their associated significance levels, resulting from the multivariate analyses of the deviation scores, that may be examined. For the iteration effect these values are $F=11.94$ ($p<0.004$) for the first offer made in a play, $F=3.92$ ($p<0.062$) for the first offer made to the winning coalition, $F=1.53$ ($p<0.288$) for the first tentative coalition formed by the winning coalition, and $F=1.55$ ($p<0.284$) for the outcomes. The mean deviation scores for the first offers, collapsed over Game and Experiment, were -8.45, -3.28, -2.48, and -3.02 for iterations 1, 2, 3, and 4, respectively, showing that the first offer made the first time a game was played was considerably more egalitarian than first offers made on the second, third, and fourth iterations. Exactly the same type of effect was found for the first offer made to the ultimately winning coalition (the mean deviation scores were -6.73, -1.90, -1.41, and -1.73 in this case), although its magnitude was smaller and its significance only marginal. The iteration effect vanished altogether when the first accepted offers by the ultimately winning coalition and the outcomes were considered. The results of the latter two analyses show that both within play and between play learning occur with regard to the disbursement of rewards. that most of it takes place from the time the first offer is made to the time the first tentative coalition is formed by members of the winning coalition, and that in both cases the overall trend is toward the quota prediction. This observed move towards the quota is a result of the negotiation processes, and, more directly, a function of the ability of the players in the coalitions other than the one that has been tentatively accepted making their strengths felt, not only for their own coalitions, but the others as well.

Despite the progressively decreasing deviations between the quota predictions on one hand, and the proposed, tentatively accepted, and ratified $p.c.$'s on the other hand, the significant game effect, unlike the iteration effect, did not vanish. On the contrary, inspection of the multivariate F values for the Game effect in the analyses of the deviation scores shows that they increased as the play progressed. The F values were $F=6.78$ ($p<0.021$) for the first offer made in a play, $F=8.61$ ($p<0.012$) for the first offer made to the winning coalition, $F=8.08$ ($p<0.014$) for the first tentative coalition formed by the winning coalition, and $F=14.54$ ($p<0.003$) for the outcomes. As a play developed the differences among the characteristic functions assumed an increasing importance in determining the disbursement of the reward for members of a particular coalition. In addition to these allocation differences, recall that the incidence of type of final coalition differed across games (Table V) as well as in the percentages of justified and unjustified objections to tentative coalitions (see Table IX).

The most obvious explanation of these effects lies in grouping the five games into two classes. Table I shows that the difference between the two classes is directly related to the index of interplayer distance Δ discussed above. For Games I, IV, and V, this index assumes the values 0.417, 0.317, and 0.447, respectively, whereas for Games II and III the values of Δ are 0.083 and 0.137, respectively. The data may be interpreted as showing that when Δ is large, player C becomes subjectively ineffective, and the AC and BC coalitions are of little interest to the players. Coalition AB so predominates the environment in such games that the three-person game is frequently considered as rather more a two-person game. Player B, perceiving the situation more as a two-person game, would be more insistent on an equal share of the results, forcing A to receive less than his quota. However, for games in which Δ is relatively small (Games II and III), player C is a more relevant participator, the AB coalition is not predominant, and the principal question is which coalition to form. Here player A may exercise his power in the negotiations, receiving his quota and perhaps even more.

The evidence reported in Table VII is consistent with the above interpretation. Table VII shows that there are fewer messages in Games II and III than in I, IV, and V, and that they take less time to complete. As suggested above, since all the three potential coalition structures are active in Games II and III, there is a serious threat of being left out of a coalition if a player gets overly ambitious. Players in such games would not prolong negotiations for fear of disrupting possible alliances with the other players and being left outside of any coalition. On the other hand, in Games I, IV, and V, where the AB coalition is predominant (and particularly in the latter two games), the use of the AC and BC coalition is frequently, though not always, for A and B to jockey around to a final agreement. They will break agreements with each other in order to establish a later, more advantageous allocation of rewards in the same coalition, if they feel it possible. Hence, more time will be spent bargaining.

We have seen that both game and iteration effects account for the discrepancy between the prescribed quota solutions and the observed $p.c.$'s expressed in the first offers, first offers made to the winning coalition, and first tentative coalitions formed by the winning coalition. Further information shows that in addition to these two factors, the discrepansies were also affected by who originated the $p.c.$ The originator of an offer may be classified as either the stronger or the weaker player in a coalition. For the first offer made in each play, the mean deviation from quota was -1.36 for a stronger originator, whereas the weaker originator's mean deviation was -6.07. This difference is highly significant by the normal approximation to the t distribution ($z > 3$, $p < 0.001$, for 237 df.). Recalling how deviation from quota scores are computed, the results show that while the stronger player was willing to 'shave' 1.36 points on the average from his quota in his first offer, the weaker player proposed an excess of 6.07 points on the average for himself. For the first offer to the winning coalition, the mean deviations from quota were -2.59 and -3.19 for the stronger and weaker players, respectively. The difference between the means is not significant in this case ($z < 1$, for 237 df.), showing that both players are now proposing that the stronger of the two shave his quota by approximately the same amount. However, for the first accepted offers by the winning coalition, the mean deviations were -2.54 and 0.98 for the stronger and weaker players, respectively. The difference between the means is significant ($z > 3$, $p < 0.001$, for 237 df.) but in the opposite direction from the significant difference found for the first offers, indicating that in the first tentative coalitions players are now making offers more to the liking of the proposed partner. Both players of the ultimately winning coalitions shaved their quotas in their first tentative agreement, though by different amounts. Riker provided evidence showing that players shaved points from their expectations in order to buy a potential ally, and to not be left out with zero rewards. The results reported above show that only the stronger player was willing to shave his quota during the entire game. The weaker player demanded more than his quota in the first offer, then decreased the excess of points he required, and finally also shaved his quota when included in the first tentative coalition formed by the ultimately winning players.

In addition to point shaving, another reason why players did not use the quota solution exactly is a tendency to round off values in offers. Recall that the preponderance of exact quota solutions occurred in

Games I and II, where the reward is exactly divisible by 5 for each coalition. Inspection of Tables II, III, and IV shows that 182 out of 240 games played ended in an outcome where one or both players received an amount of points divisible by 5. When this round-off is added to the phenomenon of point shaving, then the results appear even closer to the quota solutions. Other inspections of the data reveal that this use of values divisible by 5 was a ubiquitous phenomenon, with no differences in its frequency of occurrence being found attributable to game, communication conditions, coalition formed, or player making the offer.

Although there were no statistically significant differences in our measures due to communication conditions, there are interesting trends that deserve comment and might be useful for future research.[3] Table XIII presents the mean deviation scores by type of coalition formed and by Experiment. It can be seen that the first experiment, in which players spoke in formalized order, with no secret offers possible, produced results least like the quota solutions and most towards the egalitarian solution. It is this first condition which is the least like ordinary bargaining, and the most like the typical social scientific traditional experiment. In real life, there are ordinarily ample opportunities for secret communications among potential members of coalitions, and the possession of such confidential information makes more salient for a particular player his strength

TABLE XIII

Mean deviation from quota of outcomes by coalition type and experiment

	Coalition AB	Coalition AC	Coalition BC
Experiment 1	−3.49	−3.35	−2.58
Experiment 2	−0.91	−1.27	−0.42
Experiment 3	−1.29	−0.61	−1.40

vis-a-vis the other players. When all communications are public it might be possible for the B and C players to form a coalition not to share their meagre rewards, but rather to force player A to give up some of his share (Maschler, 1963). When secret offers are possible, player A can use this fact to restrict his conversations with players B and C so that neither knows what A is negotiating separately with them, thus making this type of coalition formation more difficult and thereby getting player A an outcome closer to his quota.

C. *The Bargaining Process*

There are several reasons why the answer to our third question, which is concerned with the characterization of the bargaining process rather than its outcome, is necessarily incomplete. One reason has to do with the *Coalitions* program used to administer the bargaining task. Although providing a common, easy to learn language for bargaining and standardization of the negotiation moves, which considerably facilitates their analysis, it does so by forcing players to communicate with one another by using only a small set of legal keywords. Various kinds of threats, promises, bluffs, and other intricate negotiation tactics are not directly expressable by this highly restricted vocabulary. Added to it is the almost inevitable problem, common to any bargaining experiment, that a large portion of the bargaining process is implicit, especially when the bargainers are experienced, and only a systematic post-experimental interrogation of the players or a talking aloud procedure may reveal its basic structure.

These caveats taken into consideration, the analysis of the bargaining process shows that the theoretical distinction between justified and unjustified objections does have behavioral implications. Justified objections are more likely to dissolve tentative coalitions than unjustified objecions, and, more importantly, the differential effect of these two types of objections steadily increases as more experience is gained with the bargaining task. Consonant with the distinction between the two classes of games we have discussed, the results (Table IX) show that justified objections are more likely to dissolve the tentative coalitions to which they are addressed than the unjustified objections when all three coalitions are viable (Games II, III, and to some extent I) but not when coalition AB predominates the other two. It is interesting to note that when justified objections are classified by game, source of objection, and target of objection (Table XI), they conform to the distinction between the two classes of games shown by earlier results. However, when the unjustified objections are classified in the same manner, they are addressed about evenly to the stronger and weaker players, as can be seen in Table XI. This, together with the Iteration effect reported in Table X, leads to speculation that justified objections were used by players with some facility and effectiveness in the game, whereas unjustified objections were more frequently lodged by unskilled players. Further evidence supporting this conjecture comes from the data presented in Table XII, where it is shown that the objections that constituted offers near the quota solution had the highest likelihood of being successful in disrupting the tentative coalitions. Investigation of this and other related hypotheses await further analyses of the bargaining process in subsequent work.

Despite these effects of justified versus unjustified objections, the results cannot be strictly interpreted as supportive of the bargaining process model. Even in later iterations, most of the justified objections were ignored rather than serving to dissolve the tentative coalitions to which they were addressed. Two deficiencies of the bargaining process model may account for its failure. One concerns the size of the set of *p.c.*'s a player is assumed to consider before making an objection, and the other concerns the effect the 'distance' between a tentatively accepted offer and $M_1^{(i)}$ should, but does not presently have on its dissolution by a justified objection. These deficiencies have been discussed in detail elsewhere (Horowitz and Rapoport, 1974) and will not be repeated here. A detailed inspection of the game protocols suggests a third weakness. Frequently, when an $M_1^{(i)}$-stable tentative coalition was formed, say between A and B, a justified objection by A against B lodged by player C was not accepted by A, not because A considered the cost of disrupting the coalition to be prohibitively high, but because he expected C to make one or two more objections, still justified, on the following rounds of negotiations, yielding A a higher gain. It seems that the model's requirement that the first justified objection against a tentative coalition should dissolve it is too restrictive and should be relaxed.

Further research indicated by this study takes on three aspects. First, a more detailed inspection of the bargaining process, for both the present and for future data, is indicated, using modified models as suggested above. Second, there is a need to explore in more detail the various social psychological aspects of this situation, by varying in more ways the communication procedures, and perhaps by introducing experimental confederates to play prescribed strategies that represent bargaining stereotypes (i.e., Boulewarism). Such research would help tie the present paradigm more closely to the recorded observations of real bargaining situations. Third, more of the mathematical aspects of n-person games need exploring. For instance, the introduction of a non-zero *ABC* coalition allows the Shapley value to come into play. Findings supportive of one or the other of the various models could suggest underlying psychological variables that determine how individuals make decisions in negotiation environments.

University of North Carolina, Chapel Hill

NOTES

* Now at the University of Southern California.
** This research was supported in part by a PHS Grant No. MH-10006 from the National Institute of Mental Health and in part by a University Science Development Program Grant No. GU-2059 from the National Science Foundation. The authors wish to thank Bruce Taylor and Sandra G. Funk for assistance in data collection, Michael Maschler for his careful reading and insightful comments on an earlier draft of the manuscript, and Abraham D. Horowitz for assistance in data collection, data analysis, and for many valuable discussions.
[1] Ideally, each subject should play each game with a different set of coplayers. The logistics of such a design are, however, beyond the scope of present-day equipment. Nonetheless, the present procedure, where a subject is not only ignorant of the roles played by his co-players, but also which players are included in a particular game, does establish some degree of independence in the sense that the outcomes are not determined by the players' gains on previous rounds. Of course, games are not strictly independent because of all players' experience with the situation, but this is a variable we shall scrutinize below.
[2] This authors wish to thank Mark I. Appelbaum for his advice concerning the multivariate data analyses.
[3] We are indebted to Michael Maschler for pointing out these particular results. Our interpretation of them, however, only partly agrees with the one he suggested.

BIBLIOGRAPHY

Aumann, R. J. and Maschler, M., 'The Bargaining Set for Cooperative Games' in M. Dresher, L. S. Shapley, and A. W. Tucker (eds.), *Advances in Game Theory*, Princeton University Press, Princeton, N.J. 1964.

Caplow, T., 'A Theory of Coalition Formation', *American Sociological Review* **21** (1956) 489–493.

Caplow, T., 'Further Development of a Theory of Coalitions in the Triad', *American Journal of Sociology* **54** (1959) 488–493.

Caplow, T., *Two Against One: Coalitions in Triads*, Prentice-Hall, Englewood Cliffs, N.J., 1968.

Chertkoff, J. M., 'A Revision of Caplow's Coalition Theory', *Journal of Experimental Social Psychology* **3** (1967) 172–177.

Davis, M. A. and Maschler, M., 'The Kernel of a Cooperative Game', *Naval Research Logistics Quarterly* **12** (1965) 223–259.

Gamson, W. A., 'A Theory of Coalition Formation', *American Sociological Review* **26** (1961) 373–382.

Horowitz, A. D., 'The Competitive Bargaining Set for Cooperative *n*-Person Games', *Journal of Mathematical Psychology* (1973).

Horowitz, A. D. and Rapoport, Amnon, 'Test of the Kernel and Two Bargaining Set Models in Four- and Five-Person Games', this volume, p. 161.

Kahan, J. P., *Coalitions: A System of Programs for Computer-Controlled Bargaining Games: Operating Manual*, L. L. Thurstone Psychometric Laboratory Research Memorandum No. 34, Chapel Hill, N. C., 1970.

Kahan, J. P., 'Noninteraction in an Anonymous Three-Person Prisoner's Dilemma Game', *Behavioral Science* **18** (1973) 124–127.

Kahan, J. P. and Helwig, R. A., 'Coalitions: A System of Programs for Computer-Controlled Bargaining Games', *General Systems* **16** (1971) 31–41.

Komorita, S. S. and Chertkoff, J. M., 'A Bargaining Theory of Coalition Formation', *Psychological Review* **80** (1973) 149–162.

Luce, R. D. and Raiffa, H., *Games and Decisions: Introduction and Critical Survey*, Wiley, New York, 1957.

Marwell, G. and Schmitt, D., 'Cooperation in a Three-Person Prisoner's Dilemma Game', *Journal of Personality and Social Psychology* **21** (1972) 376–383.

Maschler, M., 'The Power of a Coalition', *Management Science* **10** (1963) 8–29.

Maschler, M., *Playing an n-Person Game, an Experiment*, Econometric Research Program, Research Memorandum No. 73, Princeton University, 1965.

Rapoport, A., *Two-Person Game Theory: the Essential Ideas*, The University of Michigan Press, Ann Arbor, 1966.

Rapoport, A., *n-Person Game Theory: Concepts and Applications*, The University of Michigan Press, Ann Arbor, 1970.

Rapoport, A., 'Three- and Four-Person Games', *Comparative Group Studies* **2** (1971) 191–226.

Riker, W. H., 'Bargaining in a Three-Person Game', *American Political Science Review* **61** (1967) 642–656.

Selten, R. and Schuster, K. G., 'Psychological Variables and Coalition-Forming Behavior', in K. Borch and J. Mossin (eds.), *Risk and Uncertainty*, Macmillan, London, 1968.

Shapley, L. S., 'A Value for *n*-Person Games', in H. W. Kuhn and A. W. Tucker, (eds.), *Contributions to the Theory of Games*, Vol. II, Princeton University Press, Princeton, N.J., 1953.

Simmel, G., 'The Number of Members as Determining the Sociological Form of the Group', *American Journal of Sociology* **7** (1902) 1–46.

Vinacke, W. E. and Arkoff, A., 'An Experimental Study of Coalitions in the Triad', *American Sociological Review* **22** (1957) 406–414.

Vinacke, W. E., Crowell, D. C., Dien, D., and Young, V., 'The Effect of Information about Strategy in a Three-Person Game, *Behavioral Science* **11** (1966) 180–189.

von Neumann, J. and Morgenstern, O., *Theory of Games and Economic Behavior* 2nd ed. Princeton University Press, Princeton, N.J. 1947.

ABRAHAM D. HOROWITZ* AND AMNON RAPOPORT

TEST OF THE KERNEL AND TWO BARGAINING SET MODELS IN FOUR- AND FIVE-PERSON GAMES**

ABSTRACT. Employing a computer-controlled experimental paradigm for studying coalition formation and bargaining, the present study tests three models for n-person games in characteristic function form, namely, the bargaining set and two of its subsets, the competitive bargaining set and the kernel.

Twelve groups of subjects participated in several four-person and five-person Apex games. The effects of group size, order of communication, learning, and values of the characteristic function were systematically investigated. The final outcomes reject the kernel and support the two bargaining set models; they depend upon group size and order of communication.

Models describing the bargaining process, rather than the final outcomes only, are presented, tested, and partially supported. The relationships between the final outcomes of the present study and those of previous studies of Apex games are briefly discussed.

I. INTRODUCTION

An experimental paradigm has been proposed by Kahan and Rapoport (1974) for investigating coalition formation and bargaining processes in small groups. The paradigm is based on considerations of coalition formation through the negotiated division of rewards, or values, available to each coalition that may be formed. Its orientation arises from n-person game theory (see, e.g., Luce and Raiffa, 1957; Rapoport, 1966, 1970; von Neumann and Morgenstern, 1947), in particular from that portion of the theory concerned with formalized models of conflict of interest among n players, which depend only on the respective values of the possible coalitions. Utilization of the paradigm relies heavily upon the development of the digital computer as an instrument of the psychological laboratory for conducting on-line group decision making experiments.

Whereas Kahan and Rapoport (1974) have studied coaliton formation and bargaining processes in the triad, the present study, employing their experimental paradigm, has moved a step further to the case $n \geqslant 4$. Among the various games that may be investigated, we have focused on a psychologically intriguing game, first introduced by von Neumann and Morgenstern (1947, pp. 473–503) and later explored by Davis and Maschler (1965) and Horowitz (1973). The game under consideration, called the *Apex game* by Horowitz (1971), is a cooperative n-person game, $n \geqslant 3$, in which the only coalitions assigned positive values are (i) all those coalitions which include a certain player called *Apex*, and (ii) the coalition formed by the other $n-1$ players, called *Base* players.

The apex game may be cast in terms of the characteristic function of the n-person game, a real-valued set function assigning a real number $v(X)$ to each nonempty subset X of players, where $X \subseteq N$ and $N = = \{1, 2, ..., n\}$. The value $v(X)$ measures the worth or power which the coalition X can achieve when its members act together. For example, consider an Apex game with $n=5$ in which every coalition may win c units, $c>0$. Then, assuming that A is the Apex player, and B, C, D, and E denote the four Base players, the characteristic function of this game is $v(AB) = v(AC) = v(AD) = v(AE) = v(ABC) = v(ABD) = v(ABE) = = v(ACD) = v(ACE) = v(ADE) = v(ABCD) = v(ABCE) = v(ABDE) = = v(ACDE) = v(ABCDE) = v(BCDE) = c$, and $v(X) = 0$ for any other coalition X.

The Apex's position may be compared to that of a monopolist, with the only limitation that he must find at least one ally. Only the coalition of all other players against him may defeat him (von Neumann and Morgenstern, 1947). The Base's position poses an intriguing dilemma: he must either cooperate with all other Base players, regardless of their number, or he must join the Apex and possibly some other Base players. If the first course of action is chosen, the Base player risks being frozen out of a winning coalition, if one or more Base players yield to the temptation of extra gain by forming a coalition with the Apex. On the other hand, if he chooses to negotiate with the Apex, the Base must consider the highly competitive environment produced by the Apex's multitude of choices in stating his demand for his share.

The central issue of the Apex game, as well as any other n-person cooperative game in characteristic function form, has been stated succinctly by Anatol Rapoport: "Given a particular coalition structure, how will the payoffs accruing to each coalition be apportioned among its members?" (1971, p. 194) Answers to this question may be derived from some of the models proposed for n-person cooperative games in characteristic function form, namely, von Neumann and Morgenstern's *solution* (1947), Shapley's *value* (1953) and its modifications, Aumann and Maschler's *bargaining set* (1964), the *kernel* of Davis and Maschler (1965), and Horowitz's *competitive bargaining set* (1973) especially developed for Apex games. In attempting to test these models, we have discarded von Neumann and Morgenstern's solution because of the infinite number of imputations it contains, and because it is formally limited to the coalition of all n players, the *grand* coalition. Shapley's value has been discarded too because it is limited to the grand coalition. We have been left, then, with three models to test, the bargaining set, the kernel, and the competitive bargaining set.

A. *The Competitive Bargaining Set*

The basic concepts of the bargaining set and kernel models have been presented and discussed by Aumann and Maschler (1964), Davis and Maschler (1965), Horowitz (1973), Kahan and Rapoport (1974), Rapoport (1970), and will not be repeated here. Since familiarity with the competitive bargaining set model (Horowitz, 1973) cannot be assumed, its principal ideas are presented below.

Consider a cooperative n-person game in characteristic function form, which consists of a set $N = \{1, 2, ..., n\}$ of n players along with a characteristic function v, assigning the real number $v(X)$ to each nonempty subset X of players, called the coalition X. $v(X)$ is assumed to satisfy

(i) $v(X) \geq 0$ for each coalition X,
(ii) $v(\{i\}) = 0$ for each one-person coalition.

Let X be an m-partition of N satisfying

$$X_j \cap X_k = \phi, \quad \text{if } j \neq k, \quad \text{and} \quad \bigcup_{j=1}^{m} X_j = N.$$

An *outcome* of the game is represented by a *payoff configuration* (*p.c.*)

$$(\mathbf{x}; \mathsf{X}) = (x_1, x_2, ..., x_n; X_1, X_2, ..., X_m)$$

where $\mathbf{x} = (x_1, x_2, ..., x_n)$ is an n-dimensional real vector, called the *payoff vector*, representing the realizable distributions of wealth among the n players, x_i is the amount received by player i in the distribution $\mathbf{x}$, and $\mathsf{X} = \{X_1, X_2, ..., X_m\}$ represents the coalition structure which was actually formed.

A *p.c.* is assumed to satisfy individual rationality, i.e.,

$x_i \geq 0$ for all $i \in N$.

It is further assumed that

$$\sum_{i \in X_j} x_i = v(X_j), \quad \text{for} \quad j = 1, 2, ..., m.$$

A third, key, assumption is that every pair of players who are members of the same coalition ought to be in equilibrium. The concept of equilibrium is crucial to the competitive bargaining set and is defined in terms of the notions of multi-objection and counter-multi-objection.

Following the notation of Davis and Maschler (1967), let $(\mathbf{x}; \mathbf{X})$ be a p.c. and k and l be two members of the same coalition X_j, where $X_j \in \mathbf{X}$. Let $Y_1, Y_2, ..., Y_t$ be t distinct coalitions, $t \geq 1$, and let $\mathbf{y}^{(1)}, \mathbf{y}^{(2)}, ..., \mathbf{y}^{(t)}$ be the associated payoff vectors. The set

$$\{(\mathbf{y}^{(1)}; Y_1), (\mathbf{y}^{(2)}; Y_2), ..., (\mathbf{y}^{(t)}; Y_t)\}$$

is called a *multi-objection* of player k against l with respect to the p.c. $(\mathbf{x}; \mathbf{X})$, if

(i) $k \in Y_g, \quad l \notin Y_g, \quad k, l \in X_j, \quad g = 1, 2, ..., t,$

(ii) $\sum_{i \in Y_g} y_i^{(g)} = v(Y_g), \quad g = 1, 2, ..., t,$

(iii) $y_k^{(g)} > x_k, \quad y_i^{(g)} \geq x_i \quad \text{for all} \quad i \in Y_g, \quad g = 1, 2, ..., t.$

In his multi-objection, player k claims that he can gain more in any new coalition, Y_g, that he may form without the consent of player l, and that the new coalition is reasonable because the partners of k in Y_g gain at least what they gained in $(\mathbf{x}; \mathbf{X})$.

When $t = 1$, player k may threaten l through a single coalition only. In terms of the terminology of the bargaining set model, k is said to have an *objection* against l. Thus, when $t = 1$, the competitive bargaining set notion of multi-objection reduces to an ordinary objection.

For a coalition Z and a payoff vector $\mathbf{z}$ to its members, the pair $(\mathbf{z}; Z)$ is called a *counter-multi-objection* to k's multi-objection against l, if

(i) $l \in Z, \quad k \notin Z,$

(ii) $\sum_{i \in Z} z_i = v(Z),$

(iii) $z_i \geq x_i \quad \text{for all} \quad i \in Z,$

(iv) $z_i \geq y_i^{(g)} \quad \text{for all} \quad i \in (Z \cap Y_g), \quad g = 1, 2, ..., t.$

In lodging his counter-multi-objection, player l claims that he can protect his share in $(\mathbf{x}; \mathbf{X})$ by giving his partners in Z at least what they had before in $(\mathbf{x}; \mathbf{X})$, without k's consent. Moreover, if any of l's partners in Z is included in the multi-objection of k against l he would gain at least what he had gained before.

If a counter-multi-objection intersects no more than one of the coalitions $Y_1, Y_2, ..., Y_t$, it reduces to the notion of *counter-objection* of the bargaining set model.

A multi-objection to a *p.c.* is *justified* if no counter-multi-objection to it exists; otherwise it is *unjustified*. An individually rational *p.c.* is said to be $H_1^{(i)}$-*stable* if for each multi-objection of k against l in $(\mathbf{x}; \mathbf{X})$ there exists a counter-multi-objection of l against k. The set of all $H_1^{(i)}$-stable *p.c.*'s (which may be empty) is called the *competitive bargaining set* and denoted by $H_1^{(i)}$. The superscript of H indicates individual rationality and the subscript denotes that objections are made by *one* player against *one* player at a time. Both constraints may be replaced, resulting in more severe requirements of stability.

For example, individual rationality may be replaced by coalitional rationality. Formally, a *p.c.* is said to satisfy *coalitional rationality* if for any coalition W, $W \subseteq X_j, j = 1, 2, ..., m$,

$$\sum_{i \in W} x_i \geqslant v(W).$$

The set of all coalitionally rational *p.c.*'s in which no player has a justified multi-objection against any other member of the same coalition is called the *competitive bargaining set* H_1. It can be shown that $H_1 \subseteq H_1^{(i)}$. For the Apex games considered in the present paper $H_1 = H_1^{(i)}$.

B. Comparison of the Models

The bargaining set is a special case of the competitive bargaining set when $t = 1$. In particular, the *bargaining set* $M_1^{(i)}$ is defined to be the set of all individually rational *p.c.*'s in which no player has a justified objection against any other member of the same coalition. Replacing individual by coalitional rationality yields the bargaining set M_1. For the Apex games considered in the present study $M_1 = M_1^{(i)}$.

Horowitz (1973) proved that $H_1^{(i)} \subseteq M_1^{(i)}$. The proof that the kernel of the n-person cooperative game in characteristic function form, denoted by K, is also contained in $M_1^{(i)}$ is given in Davis and Maschler (1965).

It is worth mentioning that the concepts of objection and counter-objection underlying the bargaining set theory as well as the concepts of multi-objection and counter-multi-objection involve only ordinal comparisons of utilities. Although the kernel, as originally formulated by Davis and Maschler (1965), requires inter-personal comparability of utilities, Maschler, Peleg, and Shapley (1970) have recently suggested an alternative interpretation of the kernel which avoids such comparisons. The assumption of ordinal preferences, which is common to all three models tested in the present paper, considerably enhances their attraction to social scientists interested in coalition formation.

In comparing the bargaining set and the competitive bargaining set models it is noted that the degree of stability and competitiveness in bargaining behavior implied by the former model results from its assumption that only a single objection may be expressed at any one time. Underlying the competitive bargaining set model is another assumption, namely, that since threats or offers are often tacit (see, e.g., Schelling, 1960), their number should not be restricted to one. Rather, threats or offers are assumed to be perceived and considered simultaneously even though their simultaneous implementation may be impossible. The resulting outcome for Apex games is that the competitive bargaining set theory yields a unique solution located at one extreme of the continuum of solutions prescribed by the bargaining set model. The other extreme point is the kernel. For a detailed comparison of the three models with regard to Apex games see Horowitz (1973).

We are faced then with testing three models for the Apex game, $M_1^{(i)}$ and its two extreme points, $H_1^{(i)}$ and K. We expect that within the continuum of solutions prescribed by the bargaining set the actual disbursement of payoffs for a given coalition will depend on the game environment. The game environment consists of some ill-defined variables, which are psychologically important though presently not incorporated in the game theoretic formulation. Included among these variables are the determinateness to form a coalition, the degree of desired stability, and the degree of competitiveness. All of these variables, which cannot be assumed to be mutually independent, may be expected to depend upon the structure of the characteristic function and the type of communication allowed. Horowitz (1973) suggested that as the degree of determinateness to form coalitions gets stronger, the degree of competitiveness is reduced, or the stability requirements are relaxed, the final outcomes will approach the kernel's predictions. And conversely, as the players, become more competitive or if they are motivated by a desire for stronger stability the final outcomes will approach the *p.c.*'s prescribed by the competitive bargaining set.

C. *An Example*

Predictions derived from the three models may best be demonstrated by an example. Consider a five-person Apex game with the characteristic function

$$v(AB) = v(AC) = v(AD) = v(AE) = v(BCDE) = 100,$$

and $v(X)=0$ for any other coalition X. This particular Apex game, similar to the games investigated in the present study, is not a full Apex game, as any coalition with an Apex player and two or more Base players is assigned the value 0. This, however, does not affect the predictions of the models concerning the division of the payoff for a two-player coalition.

Since the four two-person coalitions which are assigned 100 points each are symmetric, it is sufficient to consider coalition AB. For this coalition there exists a unique $H_1^{(i)}$-stable *p.c.*, namely,

$$(\mathbf{x}; \mathbf{X}) = (75, 25, 0, 0, 0; AB, C, D, E),$$

as can easily be shown. A multi-objection of A against B with respect to the *p.c.* $(\mathbf{x}; \mathbf{X})$ is

$$\{(y_A^{(1)}, y_C^{(1)}; AC), (y_A^{(2)}, y_D^{(2)}; AD), (y_A^{(3)}, y_E^{(3)}; AE)\}$$

where $y_A^{(g)} > 75$, $g=1, 2, 3$, and therefore $y_C^{(1)}, y_D^{(2)}, y_E^{(3)} < 25$, since $v(AC) = v(AD) = v(AE) = 100$. Player B can respond to A's multi-objection by a counter-multi-objection

$$(25, 25, 25, 25; BCDE).$$

Note also that any single objection by B against A with respect to the *p.c.* $(\mathbf{x}; \mathbf{X})$ can be countered by the latter player. Also, it is seen that for any $\delta > 0$ the *p.c.*

$$(75 - \delta, \ 25 + \delta, \ 0, 0, 0; AB, C, D, E)$$

is not $H_1^{(i)}$-stable, since a multi-objection from A against B in which

$y_C^{(1)}, y_D^{(2)}, y_E^{(3)} > 25$ is justified, i.e., it cannot be met by a counter-multi-objection of B.

It can be shown that the only p.c., given the coalition AB, which is contained in the kernel of the Apex game, is

$$(50, 50, 0, 0, 0; AB, C, D, E).$$

And finally, the only p.c.'s with the coalition structure (AB, C, D, E) for which no player has an $M_1^{(i)}$-justified objection against any other consist of the continuum

$$(50 \leqslant x_A \leqslant 75, \quad 25 \leqslant x_B \leqslant 50, 0, 0, 0; AB, C, D, E),$$

where $x_A + x_B = v(AB) = 100$.

For the coalition of the four Base players, all three models predict a unique p.c., namely,

$$(0, 25, 25, 25, 25; A, BCDE),$$

as can be easily verified.

Employing the experimental paradigm of Kahan and Rapoport (1974), one of the purposes of the present experiment is to test the three models in Apex games. An additional, equally important, purpose is to develop and test models accounting not only for the final outcomes but also for the bargaining process. Additionally, the present experiment looked at the effects of group size, order of communication among the n players, practice, and ratio of the value of the Apex coalition to the Base coalition on the final outcomes of the game and the bargaining process.

II. METHOD

A. Subjects

Sixty undergraduate male students at the University of North Carolina participated in the experiment. They were recruited by advertisements in the student newspaper which promised financial reward. The subjects were divided into 12 groups of five subjects, each group participating in two three-hour sessions.

B. Design

A $2 \times 2 \times 2 \times 2$ factorial design was employed, with repeated measures on two of the four factors. One factor, O, was the order of communication, in which the Apex either communicated before all the Base players (O_1) or after them (O_2). A second factor, V, concerned the value of a coalition between the Apex player and a single Base player, hereafter called the *Apex coalition*. This value was either 72 (V_1) or 108 (V_2). The value of the coalition of all $n-1$ Base players, hereafter called the *Base coalition*, was always 72 points. The third factor, N, was the size of the group, either a quartet (condition N_1, an Apex plus three Base players) or a quintet (condition N_2, an Apex plus four Base players). Each of the eight games defined by the Cartesian product of the three factors N, V, and O was played twice to yield a fourth factor of runs, R, with first (R_1) and second (R_2) plays as levels.

Three groups of five subjects each were assigned to each of the four $O \times V$ combinations. Repeated measures on factors N and R were employed. Each group participated in two three-hour sessions, which typically took place within a single week. The first was a practice session. In the second session each group played two four- and two five-person Apex games, with a single passive player in the former case, who could observe the bargaining but neither send nor receive any messages. Players were labelled A, B, C, D, and E, and were required to send typed messages in alphabetical order. The Apex was therefore player A in condition O_1 and player E in condition O_2. The design of the experiment as well as the characteristic functions of the game are presented in Table I.

TABLE I
Research design

Order	Value	Groups	Size	Characteristic Function
O_1	V_1	1, 2, 3	N_1	$v(AB)=v(AC)=v(AD)=v(BCD)=72$
			N_2	$v(AB)=v(AC)=v(AD)=v(AE)=v(BCDE)=72$
	V_2	4, 5, 6	N_1	$v(AB)=v(AC)=v(AD)=108, v(BCD)=72$
			N_2	$v(AB)=v(AC)=v(AD)=v(AE)=108, v(BCDE)=72$
O_2	V_1	7, 8, 9	N_1	$v(EB)=v(EC)=v(ED)=v(BCD)=72$
			N_2	$v(EA)=v(EB)=v(EC)=v(ED)=v(ABCD)=72$
	V_2	10, 11, 12	N_1	$v(EB)=v(EC)=v(ED)=108, v(BCD)=72$
			N_2	$v(EA)=v(EB)=v(EC)=v(ED)=108, v(ABCD)=72$

Note. The characteristic functions are the same in the two levels of factor R.

C. Procedure

The first session started by having the subjects of each group read a set of instructions. Since the instructions are given in Horowitz (1971) and are also summarized in Kahan and Rapoport (1974), they will not be presented here. Essentially, they present the bargaining game as a three-stage process, consisting of an offer stage, in which the potentials of various coalitions may be explored, an acceptance stage, in which a particular $p.c.$ is seriously considered, and a ratification stage, in which the agreement on a division of value becomes binding. Communication takes place through the use of six keywords, *offer, accept, reject, ratify, pass,* and *solo*, allowing players to propose various $p.c.$'s, accept, reject or ratify them, make no communication, or withdraw from the game.

The written instructions were followed by a verbal explanation, in which the experimenter reiterated the main rules of the bargaining game. Then each subject entered a separate cubicle containing a teletypewriter connected to a PDP-8 computer to play two example games. While playing the two games under the experimenter's supervision, the subjects were encouraged to ask questions about the rules of the game and the operation of the teletypewriter and to employ all the options provided by the computer program. At the end of the first session they were told that the number of games to be played in the second session was fixed. These additional instructions were provided to discourage the subjects from fast bargaining in order to increase the number of games played and, consequently, the amount of money earned.

Subjects returned a few days later for the experimental session, entered their respective cubicles without communicating with one another, and started immediately to play. The order of the games was randomized for each group, and the Apex role was not assigned more than once to a given subject. Role were reassigned for each game to prevent sequential effects between games and to assure that experimental effects would be attributed to role and not to bargaining strategies of individual players. An interrogation of the subjects following the second session revealed that subjects did not form any hypotheses about roles assigned to the players in successive games, nor could they successfully guess the identity of the other players in a particular game.

At the end of the second session, each subject was paid $4.50 for participation in the first three-hour session, plus 5 cents per point earned in the second session, plus a fixed sum of 75 cents per hour in the second session.

The experiment was administered with a set of PDP-8, on-line, computer programs called *coalitions*. A non-technical brief description of the main program is provided by Kahan and Rapoport (1974). For a complete, technical description see Kahan and Helwig (1971).

III. RESULTS

The basic data consist of the typed messages sent during the experimental session, starting with the first *offer* or *pass* and ending with the last *ratify*. Conceptually, it is convenient to analyze these data in terms of (i) the final outcomes, (ii) the initial phase of negotiations, and (iii) the bargaining process. To simplify the ensuing presentation of the results, a system of terminology and notation is first presented. Let $G_{g,r,n}$ denote a game, where g, $g = 1, 2, ..., 12$, is the group number, r, $r = 1, 2$, is the run number, and n, $n = 4, 5$, is the group size. For example, $G_{7,2,4}$ denotes the game played by group 7 in quartet form (condition N_1) on the second run (condition R_2). An offer is a *p.c.*. When accepted by all its members it is called a *tentative coalition*. The final tentative coalition in a game is called the *ratified coalition*. A tentative coalition is *dissolved* when one of its members rejects it explicitly, or, equivalently, enters another tentative coalition. A Base player included in a ratified coalition is a *Base winner*, otherwise a *Base loser*.

The main dependent variable is the number of points, x_β, allocated to a Base player (or to all the Base players in the case of a Base coalition) in an offer, tentative coalition, or ratified coalition. Since the value of the Apex coalition, as given by the characteristic function, is known (either 72 or 108), the Apex's share, x_α, may be obtained by subtraction.

A. *Final Outcomes*

The final outcomes of the 24 quintet and 24 quartet games are presented in Tables II and III, respectively. Table II shows that in 23 of 24 quintet games an Apex coalition was formed, yielding the Base winner a mean of 20.1 points. The range of the Base winner's payoffs was from 18 to 25. The payoffs x_β predicted by the bargaining set model fall in the closed interval

$18 \leqslant x_\beta \leqslant 36$ for both conditions V_1 and V_2. The competitive bargaining set lies at the one extreme of 18, and the kernel at the other extreme, 36. All the final outcomes fall in the bargaining set M_1. In particular, they provide strong support for the competitive bargaining set compared to the kernel.

An Apex coalition was formed in 22 of the 24 quartet games yielding the Base winner a mean of 25.8 points. The range of the Base winner's payoffs was from 18 to 36. The predicted payoffs in M_1, H_1, and K for the Base winner are $24 \leqslant x_\beta \leqslant 36$, $x_\beta = 24$, and $x_\beta = 36$, respectively. As in condition N_2, the mean final outcome supports the competitive bargaining set relative to the kernel. The bargaining set is supported by 18 of the 22 final Apex coalition outcomes. Both M_1 and H_1 are also supported by the results of the two Base coalitions that were formed in games $G_{7,2,4}$ and $G_{8,1,4}$.

To assess the effects of the four experimental conditions on the final outcomes, a $2 \times 2 \times 2 \times 2$ analysis of variance with repeated measures on

TABLE II
Final outcomes of quintet games

			O_1			O_2		
	g	r	x_α	x_β	g	r	x_α	x_β
V_1	1	1	54	18	7	1	47	25
		2	54	18		2	54	18
	2	1	53	19	8	1	0	a
		2	52	20		2	47	25
	3	1	50	22	9	1	50	22
		2	53	19		2	54	18
V_2	4	1	90	18	10	1	86	22
		2	90	18		2	83	25
	5	1	90	18	11	1	83	25
		2	90	18		2	89	19
	6	1	90	18	12	1	89	19
		2	88	20		2	90	18
	Mean			18.8	Mean			21.5
	S.D.			1.3	S.D.			3.1

[a] Ratified coalition: (9, 21, 21, 21; *ABCD*)

TABLE III
Final outcomes of quartet games

				O_1				O_2		
		g	r	x_α	x_β	g	r	x_α	x_β
V_1	1	1	48	24	7	1	36	36
		2	52	20		2	0	a
	2	1	47	25	8	1	0	a
		2	48	24		2	42	30
	3	1	47	25	9	1	40	32
		2	47	25		2	40	32
V_2	4	1	90	18	10	1	80	28
		2	90	18		2	78	30
	5	1	83	25	11	1	80	28
		2	83	25		2	80	28
	6	1	86	22	12	1	84	24
		2	84	24		2	83	25
	Mean				22.9	Mean			29.3
	S.D.				2.7	S.D.			3.5

[a] Ratified coalition: (24, 24, 24; BCD)

factors N and R (employing a multivariate approach) was conducted on the Base winner payoffs presented in Tables II and III. The significant group size effect ($F=53.4$, $p<0.001$) is not predicted by the kernel model. The competitive bargaining set model, however, predicts a difference of 6 points between quartet and quintet games for the Base winner's payoff; the observed mean difference was 5.7.

The second significant main effect was due to factor O ($F=10.2$, $p<0.02$), with the Base player winning significantly more in condition O_2 than in O_1. This effect is inconsistent with all three models, since none of them incorporates any consideration of order of communication. The other two main effects, V and R, did not contribute significantly to the final outcomes.

The only significant interaction was the two-way interaction $O \times N$ ($F=6.5$, $p<0.05$), which accounts for the different effects of order of communication in quartet and quintet games. This interaction is related to the theoretical predictions in the following way. While outcomes in the quintet games supported the competitive bargaining set, the outcomes of quartet games supported it only when the Apex player communicated first. The means of the quartet games were 22.9 and 29.3 in conditions O_1 and O_2, respectively. A 0.99 confidence interval computed for condition O_1 yielded the range $20.9 \leqslant x_\beta \leqslant 24.9$, excluding most of the continuum of the bargaining set and including H_1. A similar confidence interval in condition O_2 was $26.4 \leqslant x_\beta \leqslant 32.2$, covering about two thirds of M_1 and excluding H_1 as well as K.

B. *The Initial Phase of Negotiations*

The initial phase of negotiations is defined here as the first two rounds of communication, i.e., either the first eight or ten messages in conditions N_1 and N_2, respectively. It was extensively analyzed mainly for two reasons. First, the initial tentative coalition was formed during the first two rounds of negotiations in 47 of 48 games. Secondly, 25 of the 48 initial tentative coalitions were ratified, indicating that the initial phase of negotiations strongly affected the final outcomes.

The initial phase of negotiations may be divided into three parts that will be analyzed below: the initial orientation of the Base player, the relation between first offers and final outcomes, and the relations between responses to initial offers and final outcomes.

1. *Base's orientation.* On the first round of negotiations a Base player might attempt to either cooperate with the other Base players or form a coalition with the Apex player. A measure of initial orientation of the Base players is provided by the percentage of these players who addressed the Apex with an offer, or accepted an offer he made on the first round of negotiations. These measures, ranging between 0 and 100, were obtained for each group and subjected to a $2 \times 2 \times 2 \times 2$ analysis of variance (using again the multivariate approach) with repeated measures on factors N and R. The analysis yielded a significant group size effect ($F = 14.3$, $p < 0.01$). Whereas 82% of the Base players in the quintet games negotiated with the Apex on the first round of negotiations (in which each player could send only a single message), only 58% did so in the quartet games.

Another source of significant variation was the two-way interaction $N \times O$ ($F = 10.5$, $p < 0.01$). When the Apex was the first player that could send a message (condition O_1), the percentages of the Base players who negotiated with him were 75 and 77 for conditions N_1 and N_2, respectively. When the Apex was the last player to communicate (condition O_2), the respective values were 41% and 85%. The significant interaction is the same as the one found in the analysis of final outcomes. Since neither of the other main effects, V, R, or O, nor any of the interactions were significant, the results point again to the size of the group and order of communication as the two critical variables in Apex games.

2. *Initial offers.* Tables IV and V present the initial offers made in each game, the responses to them by Base or Apex on the first or second round of negotiations, and the final outcomes for conditions N_2 and N_1, respectively. For each group size results are presented separately for conditions O_1 and O_2. The initial offers and the final outcomes are stated as before in terms of x_β. The letters W and L indicate that the Base player addressed

TABLE IV

Initial offers, responses to initial offers, and final outcomes in quintet games

O_1						O_2				
g	r	Offer by Apex	Base's response	W or L	Final x_β	g	r	Demand by Base(s)	Apex's response	Final x_β
1	1	22	c.o.l.	L	18	7	1	22, 27	c.o.m. 25	25
	2	18	ign.	L	18		2	18, 26, 36	acc. 18	18
2	1	32	acc.	L	19	8	1	19, 20, 37	acc. 19	a
	2	32	acc.	L	20		2	25, 30, 32	acc. 25	25
3	1	22	acc.	W	22	9	1	20, 20, 26, 28	acc. 20	22
	2	20	c.o.l.	W	19		2	18, 18, 20, 20	acc. 18	18
4	1	18	c.o.m.	L	18	10	1	18, 20, 20, 22	acc. 22	22
	2	18	acc.	W	18		2	20, 21, 25	acc. 25	25
5	1	20	acc.	L	18	11	1	25, 25, 26, 30	acc. 25	25
	2	b	–	–	18		2	19, 23, 24, 25	acc. 19	19
6	1	30	acc.	W	18	12	1	19, 28, 30, 54	acc. 19	19
	2	20	acc.	W	20		2	14, 17, 18, 25	acc. 18	18
Mean		22.9			18.8	Mean		24.0		21.5
S.D.		5.6			1.3	S.D.		7.0		3.1

[a] Base coalition formed.
[b] pass.

TABLE V
Initial offers, responses to initial offers, and final outcomes in quartet games

O_1						O_2				
g	r	Offer by Apex	Base's response	W or L	Final x_β	g	r	Demand by Base(s)	Apex's response	Final x_β
1	1	28	c.o.l.	W	24	7	1	24	acc.	36
	2	20	acc.	W	20		2	24	acc.	a
2	1	36	acc.	L	25	8	1	–	off 25	a
	2	24	acc.	W	24		2	–	off 32	30
3	1	25	acc.	W	25	9	1	–	off 32	32
	2	25	acc.	W	25		2	15, 24, 30	acc. 15	32
4	1	18	acc.	W	18	10	1	30	off 20	28
	2	18	acc.	W	18		2	30	acc. 30	30
5	1	38	acc.	L	25	11	1	40	c.o.m. 28	28
	2	b	–	–	25		2	20, 25, 25	acc. 20	28
6	1	30	acc.	L	22	12	1	21, 24	acc. 24	24
	2	28	c.o.l.	L	24		2	20, 25	acc. 25	25
Mean		26.4			22.9			25.1		29.3
S.D.		6.6			2.7			5.8		3.5

[a] Base coalition formed
[b] pass

by the Apex in his first offer was a winner or loser in the game, respectively.

An inspection of the column 'Offer by Apex' in both tables for condition O_1 reveals a strong run effect in Apex's initial offers; Apex's offer in the second run was never larger than in the first run. This finding strongly suggests that an Apex player in the second run learned that the Base coalition was unlikely to form, therefore demanding at least what another player in the Apex role had demanded (but not necessarily had accepted) in the first run.

Tables IV and V further show that the distribution of the Apex's initial offers to Base was bimodal, with the major mode falling close to H_1, and the minor mode falling close to K. The medians of the Apex's initial offers to Base were within one or two points of H_1, and the means were 4.9 and 2.4 points higher than H_1 for quintets and quartets, respectively. A second consistent trend emerged in condition O_1 when Apex's initial offer to

Base was compared to the final outcome. The winner Base's final outcome was never larger than his share in the Apex's initial offer. Stated differently, the bargaining that ensued in condition O_1 lowered x_β in the direction of H_1. Moreover, the bargaining also reduced the variability around the mean final outcome.

Tables IV and V show that the standard deviation of Apex's initial offers to Base was larger than that of the final outcomes. This effect is significant as tested by the ratio of the variance for quintets and quartets ($F=19.6$, $p<0.01$, and $F=5.8$, $p<0.01$, respectively).

The results of condition O_2, presented on the right-hand sides of Tables IV and V, are less regular and different than those of condition O_1. A comparison of the initial offers and final outcomes in Table V shows that the winner Base obtained on the average 29.3 points, 4.2 points *more* than his average initial demand. This difference between the two means was significant ($t=2.0$, $p<0.05$). Quintet games, however, did not exhibit such a significant trend. This finding may be explained in terms of the $N \times O$ interaction, which was significant in the previous analyses. Perhaps due to the relatively few initial offers made by Base to Apex in quartet games in condition O_2, the Apex reduced his demand in order to avoid a formation of the Base coalition.

3. *Responses to initial offers.* When either a Base or an Apex player was made an offer he had to select exactly one of five possible ways of responding: (i) reject the offer (rej.), (ii) ignore the offer by either passing or, if the player was Base, addressing the Base coalition (ign.), (iii) counter-offer and demand more than initially offered (c.o.m.), (iv) counter-offer and demand less than initially offered, (c.o.l.), or (v) accept the offer (acc.). An inspection of the column labeled 'Base's response' of condition O_1 in Tables IV and V shows that the Base chose the later two ways in 20 of 22 games (16 acceptances and 4 counter-offers for less points). The two single cases of 'resistance' to Apex led to the Base's elimination from the ratified coalition (games $G_{1,2,5}$ and $G_{4,1,5}$).

The particular payoff tentatively accepted by the Base is crucial for his chances to be a winner. The column 'W or L' in Tables IV and V indicates that a tentative agreement in the higher half of M_1 on the kernel's side led finally to the exclusion of the Base player from the ratified coalition. Note the agreements involving, for a Base player, 32 points in quintet games and 36, 38, and 30 points in quartet games. Recall that $K=36$ points for both quartets and quintets. On the other hand, Base players who accepted a payoff within four points of H_1 typically ended the game as winners.

It is instructive to describe two exceptions in which the Base players lost though they accepted initial offers located two points higher than H_1. In game $G_{1,1,5}$, the Base player counter-offered for less, 20 instead of 22, and the Apex accepted. Later in the game the Base did not agree to ratify the agreement, demanding 25 points instead of 20. He ended as a loser. In game $G_{5,1,5}$, the original agreement which assigned 20 points for the Base player was disrupted by the Apex. However, in the ensuing negotiations the first Base player passed at a crucial moment after a disruption of another Base coalition, and later he reentered the Base coalition possibly set as a trap by another Base player. It seems that in both games the Base players who had been addressed initially by the Apex could have won if they had been loyal to the Apex through all phases of the negotiations.

C. *The Bargaining Process*

The two bargaining sets M_1 and H_1 can be represented as sets of solutions of conjunctive-disjunctive systems of linear inequalities involving the final outcomes as unknowns. The predictions derived from these models, which have been tested above, concern only the final outcomes of the negotiations among the n players. The models are mute with respect to the characterization of the bargaining process which leads to ratification of the tentative coalitions. But, clearly, players do not solve conjunctive-disjunctive systems of linear inequalities in order to form coalitions and disburse their values. Rather, a coalition is ratified and its value is disbursed among its members after a lengthy process of negotiations involving offers, counter-offers, acceptances, rejections, and passes, which reflect only in part the threats, counter-threats, promises, bluffs, and other negotiation steps actually considered by the players. From a psychological viewpoint, it is the bargaining process with all its intricacies rather than the final outcomes which is of primary interest. We attend to an analysis of it in the present section.

The bargaining process may be modelled in several different ways. The alternative chosen here has been motivated by the success of the bargaining set M_1 and its extreme point H_1 in accounting for the final outcomes of

Apex games. If both models provide an adequate description of final outcomes in Apex games, and we are unwilling at this juncture to prefer one or the other, a model of the bargaining process should converge to either M_1 or H_1 as its final outcome.

Depending whether convergence to either M_1 or H_1 is sought, the two bargaining models described below amount to testable dynamic interpretations of the bargaining set model and the competitive bargaining set model, respectively. Both M_1 and H_1 can be described as *p.c.*'s in which every objection (appropriately defined for each of the two models) has a counter-objection (also appropriately defined for each case). If, for a given *p.c.*, a player k can sustain a justified objection against player l, a reasonable negotiation move might be for all players included in the objection to accept it. The resulting *p.c.* may be subjected to further negotiations. If the objection is unjustified, the bargaining continues.

Figure 1 diagrams the proposed structure of the bargaining process. On the first stage of the process a coalition (*p.c.*) is tentatively formed; it may be either an Apex or a Base coalition. The tentative coalition, denoted by X in Figure 1, may either result from the initial negotiations among the n players or may follow the dissolution of a previous tentative coalition. Players are assumed to search for justified objections against X. If none exists, the tentative coalition is eventually ratified. If at least one justified objection exists, an objection to the tentative coalition, denoted by Y_s in Figure 1, is expressed through an offer (*p.c.*).

The objections involve only integral units and are made one at a time, as dictated by the rules of the game. The identity of the player expressing an objection is irrelevant. Thus, for example, if a tentative coalition $X = AB$ is formed, an objection $Y_s = AC$ by A against B may be made by A through an offer to C, or by C through an offer to A. This requirement is compatible with the definition of objection (Aumann and Maschler, 1964).

If the objection Y_s is justified, the tentative coalition X is dissolved and the objection is accepted, thus resulting in a new tentative coalition ($X \leftarrow Y_s$ in Figure 1). If, however, the objection is unjustified, the tentative coalition X is retained, and a new objection against X, ($s \leftarrow s+1$ in Figure 1) is made. If a player has only unjustified objections when it is his turn to play, his new objection is necessarily unjustified. However, it is assumed that at least one of the players possessing both justified and unjustified objections against X will eventually express a justified objec-

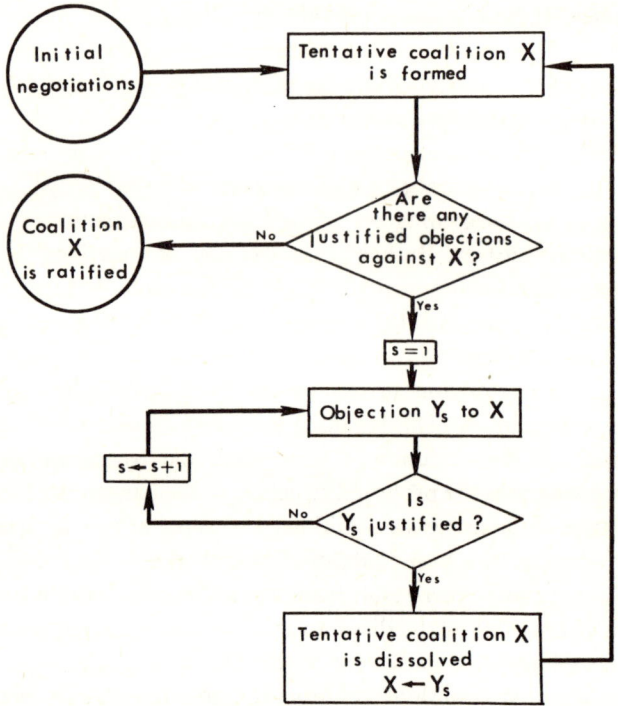

Fig. 1. A proposed bargaining process converging to the bargaining sets M_1 or H_1.

tion. The latter assumption, though admittedly strong, is required to insure convergence.

The proof for Apex games that the bargaining process described above converges to a *p.c.* in the appropriate bargaining set is too detailed to be presented here. Essentially, it is based on the idea of dividing all the *p.c.*'s in terms of x_α into five mutually exclusive and collectively exhaustive classes: (1) $0 \leqslant x_\alpha < c/(n-1)$, (2) $c/(n-1) \leqslant x_\alpha < c/2$, (3) $c/2 \leqslant x_\alpha \leqslant c(n-2)/(n-1)$, (4) $c(n-2)/(n-1) < x_\alpha \leqslant c$, (5) $x_\alpha = 0$, where c is the value of the Apex coalition. While the first four classes involve the Apex coalition only, the fifth assumes the formation of the Base coalition. The proof proceeds to show that the bargaining process described in Figure 1, assuming M_1-justified or M_1-unjustified objections, converges in a finite number of stages to the two classes comprising the bargaining set M_1, that is, either class (3) or class (5) (with an equal split of the value of the Base coalition). The proof for H_1 is quite similar. Stearns (1967) proved convergence of an entirely different transfer scheme to either M_1 or K for the general n-person game in characteristic function form.

To exemplify the bargaining process leading to M_1, consider a four-person Apex game defined for players A, B, C, D, where $v(AB) = v(AC) = = v(AD) = v(BCD) = 72$. Suppose that after a tentative Base coalition is formed with a p.c. (0, 34, 19, 19; A, BCD), the Apex player offers 21 points to C. Since this offer is an M_1-justified objection by C against B ('Base against Base'), the Base coalition is dissolved and the p.c. (51, 0, 21, 0; AC, B, D) forms after C accepts A's offer. The second iteration continues, for example, by an offer from D of 62 points to A, which is an objection by A against C (i.e., 'Apex against Base'). However, it is not M_1-justified because C can counter-object. Since the p.c. (51, 0, 21, 0; AC, B, D) is not M_1-stable, the game proceeds. For example, it may continue with a 'Base against Apex' objection by C against A (0, 22, 28, 22; A, BCD) made by B. Since it is M_1-justified it becomes the new tentative coalition. Finally, for example, an objection by B against C, $(72 - x_B, 24 \leqslant x_B \leqslant 27, 0, 0; AB, C, D)$ is both M_1-justified and stable, since C cannot counter-object. Any such objection should be ratified.

To describe the bargaining process leading to H_1, the conditions under which an offer is interpreted as a multi-objection should be specified. The bargaining process model assumes that an objection by player k against l may be viewed as a multi-objection when other objections of k against l are implied and not explicitly stated. A multi-objection may, therefore, be interpreted as a tacit threat accompanying the actual objection. Alternatively, it may be interpreted as an open offer to several potential coalitions involving player k. Thus, consider again the previous example. Suppose that players A and B agree on the p.c. (45, 27, 0, 0; AB, C, D), which is H_1-unstable. Player C may object to this p.c. by offering 46 points to A, i.e., the p.c. (46, 0, 26, 0; AC, B, D). This is an objection of the type 'Apex against Base'. The multi-objection in this case is assumed to consist of the actual objection plus the implied (tacit) objection (46, 0, 0, 26; AD, B, C). It may be interpreted as a multiple threat of A against B. Alternatively, it may be interpreted as if A is proposing 26 points to either C or D, indicating no preference between the two. Whatever is the interpretation, since the multi-objection is H_1-justified, the bargaining process continues with the dissolution of the Apex coalition, AB, until, finally, an Apex coalition is formed in which A gets 48 points.

Any of the following three events disconfirms the two bargaining process models: (i) an unstable tentative coalition not followed at least once by a justified objection, (ii) a justified objection which is ignored, i.e., does not dissolve the tentative coalition to which it is addressed, and (iii) an unjustified objection which is accepted, resulting in a new tentative coalition. The protocols of the 48 games (which appear in Horowitz, 1971) show that each of these violations occurred. With regard to the first event mentioned above, analysis of the protocols shows that of 25 M_1-unstable tentative coalitions that were formed, 17 were followed at least once by M_1-justified objections but 8 were not. Of 77 H_1-unstable tentative coalitions that were formed, 59 were followed by H_1-justified objections, but 18 were not.

A frequency analysis of objections in Apex games warrants a distinction among three types of objections: Apex against Base, Base against Apex, and Base against Base. Clearly, when an Apex coalition is tentatively formed, Apex against Base or Base against Apex types of objections may be expressed. If a Base coalition is tentatively formed, only a Base against Base type of objection may be stated.

Regardless of the type of objection, and consistent with the assumptions of the bargaining process models leading to M_1 and H_1, two cases were distinguished in the classification of the objections presented in Table VI. The first is when players stated only unjustified objections to a given tentative coalition, indicated by UJ in Table VI. The second case is when at least one of the stated objections to a tentative coalition was justified, indicated by J. In the former case, the unjustified objections were classified as either accepted or ignored, depending on whether one of them dissolved the tentative coalition or not. The same classification was maintained in the latter case, depending on whether one of the justified objections dissolved the tentative coalition or not.

The two bargaining process models leading to M_1 or to H_1 may be compared to each other only when the stated objection is of the type 'Apex against Base'. The two models yield the same predictions when a 'Base against Apex' or 'Base against Base' objection is stated. Table VI shows that there were no cases where a player stated an M_1-justified 'Apex

TABLE VI

Frequencies of tentative coalitions followed by unjustified objections only, or by at least one justified objection

Who against whom	Objection status		Objection accepted	Objection ignored
Apex against Base	M_1	Only UJ	19	50
		At least one J	0	0
	H_1	Only UJ	2	25
		At least one J	16	26
Base against Apex	M_1 and H_1	Only UJ	7	40
		At least one J	6	7
Base against Base	M_1 and H_1	Only UJ	9	2
		At least one J	4	0
Total	M_1	Only UJ	35	92
		At least one J	10	7
	H_1	Only UJ	18	67
		At least one J	26	33

against Base' objection. Of the 69 tentative coalitions for which all objections were M_1-unjustified, in 50 cases the objections were ignored. The results for model H_1 were even more impressive. Of the 27 tentative coalitions, in 25 cases the unjustified objections were ignored. However, of the 42 coalitions against which players expressed M_1-unjustified but H_1-justified objections, only 16 coalitions were dissolved. These 16 objections will be discussed in more detail below. The order of communication significant effect that was found above is reflected in the finding that 13 of these 16 cases occurred in condition O_1 and only 3 in condition O_2. And conversely, of the 26 objections that were ignored, thus discounting the bargaining process model leading to H_1, only 7 occurred in condition O_1 and 19 in condition O_2.

A frequency analysis of 'Base against Apex' objections shows that of the 13 cases in which players expressed at least one justified objection, only six were accepted. Of the 47 tentative coalitions against which all the objections made were unjustified, the objections were ignored in 40 cases. There were seven cases where the Base dissolved the Apex coalition in favor of the Base coalition for an unequal apportionment of its value. In five of these seven cases the Base disruptor ended as a loser.

There were only 15 tentative coalitions followed by 'Base against Base' objections, 13 of which were dissolved. Recalling that the Base coalition was ratified in only three of 48 games, the high percentage of accepted objections provides additional evidence to the instability of the Base coalition. The latter was dissolved by almost any objection, whether justified or not. It is worth noting the relationship between the 'Base against Base' objections that were accepted and winning or losing the game. Six of the 9 Base players who dissolved the Base coalition through an unjustified objection ended as losers, whereas all 4 players who dissolved it through a justified objection ended as Base winners.

The frequencies of objections that were either accepted or ignored, summed over the three types, are presented in the lower part of Table VI.

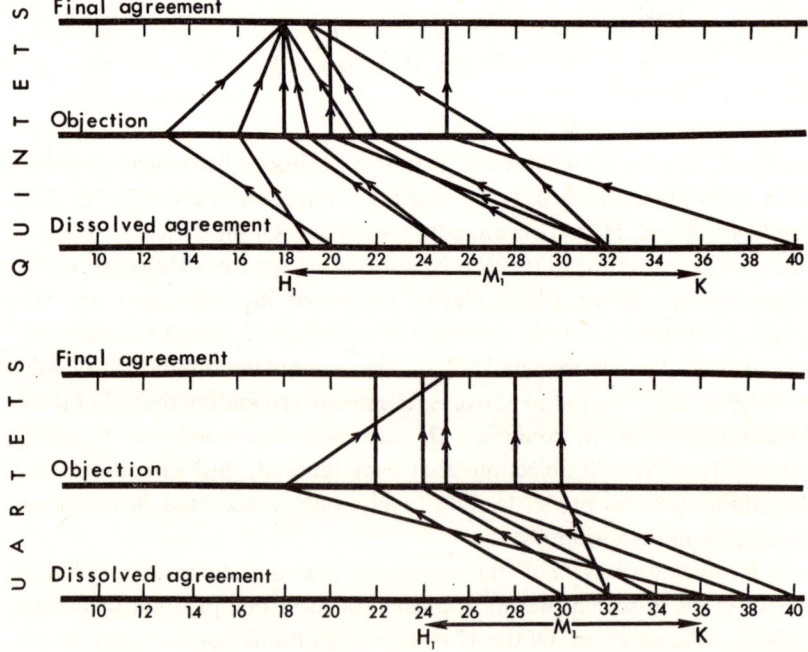

Fig. 2. Dissolutions of tentative agreements as a result of M_1-unjustified but H_1-justified 'Apex against Base' objections, in terms of Base's payoff.

The frequencies are shown separately for the two bargaining process models. The null hypothesis of no interaction between the two factors of each table was rejected ($\chi^2 = 5.44$, $p < 0.02$, for model M_1, and $\chi^2 = 7.56$, $p < 0.01$, for model H_1). Inspection of the frequency tables shows that for both models, when a player had only unjustified objections, an objection was about three times more likely to be ignored than accepted. When he had both justified and unjustified objections, a justified objection was stated and accepted in only approximately half of the cases.

As stated above, there were 16 M_1-unjustified but H_1-justified 'Apex against Base' objections dissolving the tentative coalitions, nine in condition N_2 and seven in condition N_1. These are portrayed in Figure 2. The bottom axis in each of the two halves of the figure shows the payoff to the Base player in the coalition that was dissolved. The middle axis shows the Base's payoff in the dissolving objections, and the top axis displays the payoffs to Base in the ratified coalitions. Note that in some cases more dissolutions occurred between the middle and top levels. The general pattern of results displayed in Figure 2 indicates that Apex's dissolving objections reduced the large range of payoffs to Base in the tentative Apex coalitions (practically the whole continuum of M_1) to a considerably smaller range around H_1. Three of the four games in which objections resulted in Base payoffs outside M_1 ended in M_1 as a result of later disruptions.

IV. DISCUSSION

A. *Final Outcomes*

The final outcomes of the Apex games strongly support the bargaining set model; 45 of the 48 final outcomes were included in M_1. As noted earlier, for the characteristic functions presented in Table I, M_1 comprises an interval of *p.c.*'s rather than a unique solution. One may hold the view that stronger predictions, constituting subsets of M_1, are not possible, since extra-game-theoretical considerations such as 'standards of behavior' in groups of college students, the nature of the communication channels, or the 'bargaining abilities' of the players determine particular outcomes within M_1. Von Neumann and Morgenstern (1947) presented a similar argument in defending their 'solution'. But if one is dissatisfied with the multitude of solutions in M_1 and wishes to achieve a higher level of predictability, alternative models should be investigated.

Both the competitive bargaining set and the kernel models predict unique payoff vectors for Apex games, the extreme points of M_1. The present experiment was designed to make the range between the two extreme points sufficiently large to allow statistical tests of the models even when the number of groups is relatively small. The final outcomes, however, were unambiguous, requiring no sophisticated statistical analysis of the data. Fourteen of the 48 final outcomes were included in H_1 in comparison to only a single outcome in K. Moreover, the mean final outcome for Base was within 2.1 points from H_1 for both group size conditions, whereas its distance from K was approximately five to eight times larger. The analysis of variance results, however, showed that the success of the competitive bargaining set in accounting for final outcomes depended upon the group size and order of communication. Whereas the final outcomes of the quartet games in condition O_1 and those of quintet games for both conditions of order of communication supported H_1 relative to K and most of M_1, the final outcomes of the quartet games in condition O_2 were approximately uniformly distributed within M_1.

In addition to the competitive bargaining set model, the final outcomes support any other model which makes identical predictions. There is at least one such model. Horowitz (1973) has shown that for Apex games the von Neumann and Morgenstern's main simple solution is identical to what he has called the non-trivial payoff vectors in H_1. The final outcomes, therefore, formally support the main simple solution to the extent that they support H_1.

Yet, even though both models predict the same final outcomes, they differ from each other in two ways. First, while the competitive bargaining set yields for each winning coalition in the Apex game a unique payoff vector, the main simple solution is only one of a large number of solutions, each of which containing an infinite number of imputations. A second, not less important distinction between the two theories concerns the development of a dynamic theory for the bargaining process. Stearns (1967) as well as the present study present convergence schemes leading to the bargaining sets. These dynamic interpretations of the bargaining set models can be tested by a proper analysis of the bargaining process. No such analysis seems feasible for the static theory of von Neumann and Morgenstern.

B. *Effects of Independent Variables*

From the four factors that were manipulated in the experiment, the group size and order of communication emerged as the most prominent factors. Run effects were only noted when Apex's initial offers were analyzed, whereas the value of the Apex coalition significantly affected behaviour in none of the analyses we have conducted. Factors N and O, besides significantly affecting the final outcomes, also affected the Base's bargaining behavior at the outset of the game. Since only two group size conditions were run, the results are presently not generalizable beyond $n=5$. They suggest, however, that the larger the n the less cohesive are the Base players on the first round of negotiations in Apex games. This hypothesis is testable in other experiments in which $n>5$. The results also suggest that when n is small, cohesion among the Base players increases if, rather than letting the Apex player attempt to form an Apex coalition, the communication rules present the Base players with the opportunity to briefly negotiate with one another before the Apex's intervention.

The importance of group size and order of communication is supported by two additional statistical tests, not reported above and unrelated to the final outcomes. Two $2 \times 2 \times 2 \times 2$ analyses of variance, employing a multivariate approach as before, were conducted on (i) the total number of messages sent during each game, and (ii) the number of tentative coalitions formed in each game. The only significant effects in both tests were again attributable to N, O, or their interaction.

The following picture emerges, then, regarding the bargaining behavior of the players. If the Apex player communicates first when $n=4$, he controls the game during the first phase of the experiment, making negotiations among the Base players very unlikely. But if the Apex communicates last, after the Base players, the latter are more likely to communicate with one another, thus exerting moderate pressure on the Apex, who, in turn, responds by initiating or accepting a less favorable share to insure the formation of the Apex coalition. Since almost all the first tentative coalitions were formed on the first two rounds of negotiations, and more than half of them were later ratified, presumably because the penalty for dissolving a tentative coalition was high, Apex's mean final outcome did not differ significantly from his mean demand in condition O_1, but was significantly smaller in condition O_2. Increasing the number of Base players from three to four almost completely prevented cooperation among the Base players, thus decreasing the likelihood of the formation of the Base coalition. Hence, when $n=5$, the order of communication did not affect the initial orientation of the players and, consequently, the final outcomes.

With respect to Base's behavior, the following very simple policy seems to enhance his chances to win. When Apex's initial offer to him is high relative to what he could obtain from a symmetric apportionment of the value of the Base coalition, the Base should counter-offer demanding less for himself. Otherwise, he should accept Apex's offer immediately, and remain in coalition with the Apex until ratification. The analysis of the bargaining process showed that most of the Base players who dissolved a tentative coalition with an unjustified objection ended as losers. Hence, Base's best policy is to adhere to the prescriptions of the two bargaining process models, which are the same for his bargaining behavior.

C. *The Bargaining Process*

The only data directly reflecting the bargaining process consisted of the messages typed and transmitted by the subjects. Such data provide only occasional glimpses of the bargaining process, partly because the experimental design allowed the players to communicate with only a small set of legal messages, and, more importantly, because the threats, counter-threats, promises, and other, more subtle, negotiation moves that the players might have considered, could not be reflected in the messages they sent. Additional information about the bargaining process may, perhaps, be obtained by requiring the players during the game to state the reasons for their moves and explain in as much detail as possible their thought processes. The talking aloud procedure, which has been proved useful in some problem solving studies, may provide equally fruitful information in bargaining studies.

Notwithstanding the limitations of the analysis, the results supported the two bargaining process models. In particular, the analysis showed that whether a tentative coalition was dissolved by an objection depended on whether the objection was justified. Additionally, this dependence was affected by the type of the objection. Both models were supported when the objector possessed only unjustified objections, unless the objection was of the type 'Base against Base'. The support given to the bargaining process leading to H_1 is particularly impressive since, it may be recalled, H_1 consists of only a single *p.c.*.

Both models of the bargaining process suffer from several deficiencies. The first weakness concerns the proposed test, which, if answered negatively, leads to ratification (see Figure 1). To perform this test, the players are supposed to search for justified objections against X. Although the set of objections they are assumed to consider is finite, it may be very large, making an effective search unfeasible. To allow for an effective search, the set of objections considered by each player should be restricted. A second, more serious weakness, is that the 'distance' of (x; X) from M_1 (or H_1) does not affect its dissolution by a justified objection. A more reasonable model would require the dissolution of the tentative coalition X by a justified objection Y to be probabilistically rather than deterministically determined, with the probability of dissolution increasing monotonically as the 'distance' of (x; X) from M_1 (or H_1) increases.

D. *Comparisons with Previous Studies*

The results of the present experiment may be compared to results obtained in two other experiments that employed Apex games. Maschler (1965) employed two Apex games among several three- and four-person games in characteristic function form played by Israeli high school students. The first, Game I, was defined by the characteristic function $v(AB) = v(AC) = = v(AD) = v(BC) = v(BD) = v(CD) = 50$, $v(BCD) = 111$, and $v(X) = 0$ for any other coalition X. The second, Game II, was like Game I, with the only difference being that $v(BCD) = 120$. Each of the two games was played once by each of five different quartets of players. As in the present study, the Base coalition, BCD, was formed in only one of 10 cases. In eight of the nine cases that the Apex coalition was formed, the final outcome was included in M_1 but never in K or H_1. The predicted payoffs for Game I in M_1, K, and H_1 for the Base winner were $37 \leqslant x_\beta \leqslant 43$, $x_\beta = 43$, and $x_\beta = 37$, respectively. The respective predictions for Game II are $40 \leqslant x_\beta \leqslant 47\frac{1}{2}$, $x_\beta = 47\frac{1}{2}$, and $x_\beta = 40$. The final outcomes for the Base winner were 38, 40, 40, 40, and 45 for Game I, and 41, 41, 45, and 45 for Game II.

In an experiment conducted by Selten and Schuster (1968), 12 groups of five subjects each played a five-person game with the characteristic function $v(AB) = v(AC) = v(AD) = v(AE) = v(ABC) = v(ABD) = v(ABE) =$

$= v(ACD) = v(ACE) = v(ADE) = v(ABCD) = v(ABCE) = v(ABDE) =$
$= v(ACDE) = v(ABCDE) = v(BCDE) = $ DM 40, where DM 40 equals approximately $10. The Base coalition was formed in two of 12 cases. A two-person Apex coalition was formed in eight of 12 cases, yielding the Base winner the payoffs 12, 15, 15, 15, 15, 15, 20, and 22. The predictions of M_1, K, and H_1 are $10 \leqslant x_\beta \leqslant 20$, $x_\beta = 20$, and $x_\beta = 10$, respectively. The median payoff, 15, fell in the middle of M_1. In one game a coalition among the Apex and three Base players was formed, and in another game coalition $ABCDE$ was formed for an equal split of its value.

All three experiments were designed to test the bargaining set and kernel models. The final outcomes clearly support the former model and reject the latter. Another common finding is the rarity of the Base coalition. This finding is of interest, because it seems to refute the widespread convinction, as well as the predictions of some social psychological theories, that the weak Base players are likely to unite against the strong Apex player instead of the other way around.

With regard to a comparison between M_1 and its extreme point H_1, the results of the three studies are less consistent. Whereas most of the final outcomes in the present study were distributed in one half of M_1, with an average very close to H_1, the final outcomes of the quartet games in condition O_2 and those of the experiments by Maschler (1965) and Selten and Schuster (1968) were distributed over the entire set of solutions comprising M_1, with an average very close to its middle. Because of the differences among the studies in the values of the characteristic functions, the experimental designs, and the nationality and the age of the subjects, the discrepancy between the final outcomes may be attributed to a variety of factors. In particular, and as suggested before, the discrepancy between the central tendencies of the final outcomes of the three studies in comparison to M_1 may be attributed to the difference among the studies in the form of communication. Whereas Maschler as well as Selten and Schuster allowed free, face to face negotiations, the present experiment limited the communication to a small preselected set of formal messages, without allowing the players to see or hear one another or even know the identity of the other players. Face to face contact allows the communication of intentions, gestures, and emotions, thus probably enhancing the salience of social norms of equity, decreasing competitiveness, and increasing the determinateness to form a coalition. The lack of personal contact, on the other hand, might have facilitated competitive tendencies and strengthened the desire for a stronger stability.

University of North Carolina, Chapel Hill

NOTES

* Now at General Motors Research Laboratories, Warren, Michigan, 48090.
** This research was supported in part by a PHS Grant No. MH-10006 from the National Institute of Mental Health and in part by a University Science Development Program Grant No. GU-2059 from the National Science Foundation given to the L. L. Thurstone Psychometric Laboratory at the University of North Carolina. The authors wish to thank James P. Kahan for developing the *Coalitions* computer program and for many valuable discussions, and Thomas S. Wallsten and Michael Maschler for many helpful suggestions.

BIBLIOGRAPHY

Aumann, R. J. and Maschler, M., 'The Bargaining Set for Cooperative Games', in M. Dresher, L. S. Shapley, and A. W. Tucker (eds.), *Advances in Game Theory*, Princeton University Press, Princeton, N.J., 1964.

Davis, M. and Maschler, M., 'The Kernel of a Cooperative Game', *Naval Research Logistics Quarterly* 12 (1965) 223–259.

Davis, M. and Maschler, M., 'Existence of Stable Payoff Configurations for Cooperative Games', in M. Shubik (ed.), *Essays in Mathematical Economics in Honor of O. Morgenstern*, Princeton University Press, Princeton, N.J., 1967.

Horowitz, A. D., *The Competive Bargaining Set: Development, a Test, and Comparison with the Bargaining Set and Kernel in n-Person Games'*, Research Memorandum No. 36, The L. L. Thurstone Psychometric Laboratory, University of North Carolina, 1971.

Horowitz, A. D., 'The Competitive Bargaining Set for Cooperative *n*-Person Games', *Journal of Mathematical Psychology* 10 (1973) 265–289.

Kahan, J. P. and Helwig, R. A., 'A System of Programs for Computer-Controlled Bargaining Games', *General Systems* 16 (1971) 31–41.

Kahan, J. P. and Rapoport, A., 'Test of the Bargaining Set and Kernel Models in Three-Person Games', in this volume, pp. 119–160.

Luce, R. D. and Raiffa, H., *Games and Decisions: Introduction and Critical Survey*, Wiley, N.Y. 1957.

Maschler, M., *Playing an n-Person Game, an Experiment*, Econometric Research Program. Research Memorandum No. 73, Princeton University, 1965.

Maschler, M., Peleg, B., and Shapley, L. S., *The Kernel and the Nucleolus of a Cooperative Game as Locuses in the Strong ε-Core*. RPGTME RM 60, Department of Mathematics, Hebrew University of Jerusalem, May 1970.

von Neumann, J. and Morgenstern, O., *Theory of Games and Economic Behavior*, 2nd ed., Princeton University Press, Princeton, N.J., 1947.

Rapoport, A., *Two-Person Game Theory: The Essential Ideas*, The University of Michigan Press, Ann Arbor, 1966.

Rapoport, A. *n-Person Game Theory: Concepts and Applications*, The University of Michigan Press, Ann Arbor, 1970.

Rapoport, A. 'Three- and Four-Person Games', *Comparative Group Studies* 2 (1971) 191–226.

Schelling, T. C., *The Strategy of Conflict*, Harvard University Press, Cambridge, 1960.

Selten, R., and Schuster, K. G., 'Psychological Variables and Coalition-Forming Behavior', in K. Borch and J. Mossin (eds.), *Risk and Uncertainty*, Macmillan, London, 1968.

Shapley, L. S., 'A Value for *n*-Person *Games*', in H. W. Kuhn and A. W. Tucker (eds.), *Contributions to the Theory of Games*, Vol. II. Princeton University Press, Princeton, N.J., 1953.

Stearns, R. E., *Convergent Transfer Schemes for n-Person Games*, General Electric Report No. 67-C-311, Schenectady, N.Y., 1967.

D. MARC KILGOUR

A SHAPLEY VALUE FOR COOPERATIVE GAMES WITH QUARRELLING

The Shapley value [1] has proven to be a useful evaluation of a player's strategic position in an essential cooperative game. In this paper, we shall consider two extensions of the Shapley value to games in which there is a set of players, each of whom refuses to join any coalition already containing another player in the set. Such an extension will provide information about the strategic value of cooperation in the game. For example, by subtracting a quarreller's extended Shapley value from his extended Shapley value when he is removed from the set of quarrellers, we determine the value to him of cooperation with the other quarrellers. Similarly, we can find the value to any player of cooperation among the quarrellers (if it were possible).

We begin with several definitions. The set of players is $N = \{1, 2, ..., n\}$, where $n > 2$. All sets mentioned below will be subsets of N. The set of quarrellers is Q, with $2 \leq q = |Q| < n$. Let $F = N - Q$, the set of cooperating (friendly) players. For $j \in Q$, there is exactly one possible maximal coalition to which j can belong; denote it by $C_j = F \cup j$. (Note that we do not distinguish between j and $\{j\}$.) Define $c = |C_j| = n - q + 1$. Whenever a set is called 'S', it is assumed that $s = |S|$. Whenever a player is called 'i' (respectively, 'j'), it is assumed that $i \in F$ ($j \in Q$). The group of all permutations of a set X is denoted by $\mathscr{S}(X)$.

We assume an essential game, with players N, given by its (superadditive) characteristic function v. For any player k, the 'marginal characteristic function of k', V_k, is defined by

$$V_k(S) = v(S) - v(S - k).$$

Observe that, if $k \in S$, $V_k(S) \geq v(k)$, but if $k \notin S$, $V_k(S) = 0$.

Our extensions are based upon a particular formulation of the Shapley value, when there is no quarrelling.[1] This formulation yields an obvious generalization of the Shapley value, and, more importantly, it is capable of extension when quarrelling occurs. A model of play for a game is a set of procedures governing formation of coalitions and allocation of rewards.

Anatol Rapoport (ed.), Game Theory as a Theory of Conflict Resolution, 193–206. All Rights Reserved. Copyright © 1974 by D. Reidel Publishing Company, Dordrecht-Holland

We shall always regard a player's Shapley value as his expected reward when a specified model of play has been adopted. When there is no quarrelling, this model is the following:

I. MODEL OF PLAY WITH NO QUARRELLING

The referee instructs all of the players to meet at a certain place, where the grand coalition, N, will be formed. If, when player k reaches the meeting place, the set S (with $k \notin S$) has already arrived and formed a coalition, then k joins S immediately and receives $V_k(S \cup k)$.

Suppose that each player's probability distribution of arrival time is known. Let $\sigma \in \mathscr{S}(N)$ be fixed. Then $P(\sigma)$, the probability that the grand coalition forms in the order σ, can be calculated. Define

$$S_\sigma(k) = \{h : \sigma(h) \leq \sigma(k)\},$$

for $k \in N$ and $\sigma \in \mathscr{S}(N)$. Then k's generalized Shapley value is

$$(1) \qquad \phi_k^P[v] = \sum_{\sigma \in \mathscr{S}(N)} P(\sigma) V_k(S_\sigma(k)).$$

The generalized Shapley value becomes the Shapley value when the players' distributions of arrival time are all identical. Then each possible order of coalition formation is equiprobable. In this case, Shapley has noted that[2]

$$(2) \qquad \phi_k[v] = \sum_S \gamma(n, s) V_k(S),$$

where $\gamma(n, s) = (s-1)! \, (n-s)!/n!$. We have dropped the superscript 'P' on $\phi_k[v]$ to indicate that a special choice of the probability P has been made[3].

Our first extension of the Shapley value, when quarrelling occurs, is derived from the following model:

II. FIRST MODEL OF PLAY WITH QUARRELLING

The referee selects one maximal coalition, C_j, and instructs only the members of C_j to meet at a certain place where the coalition C_j will be formed. If $k \in C_j$ and if, when k reaches the meeting place, the set S

SHAPLEY VALUE FOR GAMES WITH QUARRELLING   195

(with $S \subseteq C_j - k$) has already arrived and formed a coalition, then k joins S immediately and receives $V_k(S \cup k)$. If $k \notin C_j$, then k receives $v(k)$.

Suppose that the probability that the referee selects C_j is $P(C_j)$. For $\sigma \in \mathscr{S}(C_j)$, define $\bar{\sigma}$, the order of formation corresponding to σ, by

$$\bar{\sigma}(k) = |S_\sigma(k)|, \quad k \in C_j.$$

Suppose that each player's probability distribution of arrival time is known. Then it is possible to calculate $P(\bar{\sigma} \mid C_j)$, the probability that the coalition forms in the order $\bar{\sigma}$, given that the referee has selected C_j. Player k's generalized Shapley value is his expected reward in this model, which is

(3) $$\phi_j^P[Q, v] = P(C_j) \sum_{\sigma \in \mathscr{S}(C_j)} P(\bar{\sigma} \mid C_j) V_j(S_\sigma(j)) +$$
$$+ v(j) \sum_{k \in Q-j} P(C_k)$$

(4) $$\phi_i^P[Q, v] = \sum_{j \in Q} P(C_j) \sum_{\sigma \in \mathscr{S}(C_j)} P(\bar{\sigma} \mid C_j) V_i(S_\sigma(i)).$$

Similarly to the case when there is no quarrelling, we define the Shapley value to be the generalized Shapley value when the referee's selections are uniformly distributed and the players' distributions of arrival time are all identical:

(5) $$\phi_j[Q, v] = \frac{1}{q} \sum_{\sigma \in \mathscr{S}(C_j)} \frac{1}{c!} V_j(S_\sigma(j)) + \frac{q-1}{q} v(j) =$$
$$= \frac{1}{q} \sum_{j \in S \subseteq C_j} \gamma(c, s) V_j(S) + \frac{q-1}{q} v(j) =$$
$$= \sum_{S \cap Q = j} \frac{\gamma(c, s)}{q} V_j(S) + \frac{q-1}{q} v(j).$$

(6) $$\phi_i[Q, v] = \sum_{j \in Q} \frac{1}{q} \sum_{\sigma \in \mathscr{S}(C_j)} \frac{1}{c!} V_i(S_\sigma(i)) =$$
$$= \frac{1}{q} \sum_{j \in Q} \sum_{i \in S \subseteq C_j} \gamma(c, s) V_i(S) =$$
$$= \sum_{S \cap Q = \phi} \gamma(c, s) V_i(S) + \sum_{|S \cap Q| = 1} \frac{\gamma(c, s)}{q} V_i(S).$$

From the development of (5), it follows that

$$\sum_{S \cap Q = j} \frac{\gamma(c, s)}{q} = \frac{1}{q} \qquad (7)$$

and that (5) is an expectation in the sense that (i) the probability that j is ultimately unallied is $(q-1)/q$, and (ii) if $S \cap Q = j$, the probability that j joins $S-j$ and that ultimately C_j forms is $\gamma(c, s)/q$. (Of course, (7) is easy to prove directly.) Similarly, the sum of the coefficients (of $V_i(S)$, when $i \in S$) in (6) is unity. In fact, a direct proof that

$$\sum_{\substack{|S \cap Q|=1 \\ i \in S}} \frac{\gamma(c, s)}{q} = \sum_{\substack{S \cap Q = \phi \\ i \in S}} \gamma(c, s) = \tfrac{1}{2} \qquad (8)$$

is straightforward. Again, if $i \in S$, it is possible for i to join $S-i$ iff $|S \cap Q| \leq 1$; for each such S, the coefficient of $V_i(S)$ in (6) is the probability that i joins $S-i$.

Application of (7) to (5) and (8) to (6) shows that $\phi[Q_k, v] \geq v(k)$. But, in general, no inequality can hold between $\phi_k[Q, v]$ and $\phi_k[v]$. This is demonstrated by the example below. In particular, the example shows that it is sometimes possible for a player to increase his Shapley value by quarrelling.

EXAMPLE. Let $N = \{1, 2, 3\}$ and $Q = \{1, 2\}$. Let v be in 0−1 normalization, with $v(1, 2) = a$, $v(2, 3) = b$, and $v(1, 3) = c$, where $0 \leq a, b, c \leq 1$. Simple calculations show that

$$\phi_1[v] = \frac{a + c + 2(1 - b)}{6} \qquad \phi_1[Q, v] = \frac{c}{4}$$

$$\phi_3[v] = \frac{b + c + 2(1 - a)}{6} \qquad \phi_3[Q, v] = \frac{b + c}{4}.$$

Therefore, $\phi_1[Q, v] > \phi_1[v]$ iff

$$c > 2a + 4(1 - b). \qquad (9)$$

Roughly, (9) holds when b and c are large and a is small; we can say that 1 gains by quarrelling when 3 is strong and 1 and 2 are weak.

Similarly, $\phi_3[Q, v] > \phi_3[v]$ iff

$$b + c > 4(1-a),$$

so that, roughly, 3 gains from the quarrel of 1 and 2 when 1 and 2 are very strong players.

Now, returning to the general case, we consider the question of when a quarreller gains from the quarrel. In analogy to (9), we have from (2) and (5) that $\phi_j[Q, v] > \phi_j[v]$ iff

(10) $$\sum_{S \cap Q \ni j} \gamma(n, s) V_j(S) - \frac{q-1}{q} v(j) +$$

$$+ \sum_{S \cap Q = j} \left[\gamma(n, s) - \frac{\gamma(c, s)}{q} \right] V_j(S) < 0.$$

The following lemma provides information about the coefficients in (10):

LEMMA I. $\sum_{S \cap Q = j} \gamma(n, s) = 1/q$.
Proof. Observe that

$$\sum_{S \cap Q = j} \gamma(n, s) = \sum_{s=1}^{c} \binom{c-1}{s-1} \gamma(n, s) = \frac{(c-1)!}{n!} \sum_{r=0}^{w} \frac{(t+r)!}{r!},$$

where $w = c - 1$ and $t = n - c$. Therefore the lemma is true provided that

(11) $$\sum_{r=0}^{w} \frac{(t+r)!}{r!} = \frac{(w+t+1)!}{w!(t+1)}$$

whenever $t \geq 0$ and $w \geq 0$. It is easy to verify (11) by induction on w.

For any player k, it is clear that

(12) $$\sum_{k \in S} \gamma(n, s) = 1.$$

The combination of (12) with Lemma I shows that

(13) $$\sum_{S \cap Q \ni j} \gamma(n, s) = \frac{q-1}{q}.$$

Since v is superadditive, it follows that the first two terms of (10), taken together, are non-negative. Therefore, if j gains from the quarrel, the third term of (10) must be negative. Define

$$\alpha(n, c, s) = \gamma(n, s) - \frac{\gamma(c, s)}{q}.$$

From (7) and Lemma I,

$$\sum_{S \cap Q = j} \alpha(n, c, s) = 0.$$

The next lemma describes the behaviour of α as s varies.

LEMMA II. *There is a real number s^*, satisfying $1 < s^* < c$, such that $s < s^*$ implies $\alpha(n, c, s) > 0$, and $s > s^*$ implies $\alpha(n, c, s) < 0$.*

Proof. First, it is easy to check that

$$\alpha(n, c, 1) = \frac{1}{n} - \frac{1}{cq} = \frac{c(n + 1 - c) - n}{ncq} > 0,$$

Secondly,

$$\alpha(n, c, c) = \frac{(c-1)!(n-c)!}{n!} - \frac{1}{cq} < 0$$

$$\Leftrightarrow \binom{n}{c} > q.$$

Again, since $2 \leq c \leq n-1$, $\binom{n}{c} \geq n > q$, so $\alpha(n, c, c) < 0$. Therefore, $\alpha(n, c, s)$ changes sign at least once in the interval $1 \leq s \leq c$. To complete the proof, it suffices to show that $\alpha(n, c, s)$ changes sign at most once in this interval. Now

$$\alpha(n, c, s) \gtreqless 0 \quad \text{according as}$$

$$q \gtreqless \frac{(c-s)! \, n!}{c! \, (n-s)!} = \prod_{r=1}^{s} \left(\frac{n-s+r}{c-s+r} \right) = \beta(s).$$

If $1 \leq s \leq c-1$, then $\beta(s+1)/\beta(s) = (n-s)/(c-s) > 1$, so that β is an increasing function of s, proving the lemma.

Applying Lemma II and our previous observations to (10), we can say, roughly, that j gains from the quarrel when $V_j(S)$ tends to assume large values when $S \cap Q = j$ and s is large, relative to its values when $S \cap Q = j$ and s is small, and when $S \cap Q \supset j$.

A similar analysis can be applied to the effect of the quarrel on the Shapley value of player i, a non-quarreller. It shows that, roughly speaking, i gains from the quarrel when, for $i \in S$, $V_i(S)$ tends to assume large values when $S \cap Q = \phi$ and when $|S \cap Q| = 1$ and s is large, relative to its values when $|S \cap Q| = 1$ and s is small, and when $|S \cap Q| \geq 2$. It is instructive to compare these results with the observations concerning the situations in which 1 and 3 gain, in the example above.

Our second extension of the Shapley value, when quarrelling occurs, is derived from the following model of play.

III. SECOND MODEL OF PLAY WITH QUARRELLING

The referee instructs all of the players to meet at a certain place, where some maximal coalition will form. If, when player k reaches the meeting place, the set S (with $k \notin S$) has already arrived and formed a coalition, then k joins S immediately and receives $V_k(S \cup k)$ provided that either $k \in F$ or $S \cap Q = \emptyset$. If $k \in Q$ and $S \cap Q \neq \emptyset$, then k leaves immediately and remains unallied, receiving $v(k)$.

Suppose that each player's probability distribution of arrival time is known. Let $j \in Q$ and $\sigma \in \mathscr{S}(C_j)$. It is easy to calculate $P(C_j)$, as it is simply the probability that j is the first member of Q to arrive. The joint probability $P(C_j, \bar{\sigma})$ can also be calculated, and from it, the conditional $P(\bar{\sigma} \mid C_j)$ obtained. The generalized Shapley value for a player is his expected gain in this model of play, which is given, as in the first model, by (3) and (4).

Now assume that the players' distributions of arrival time are all identical. The distinction between the first and second models of play will now become apparent. It follows that now we have

$$(14) \quad P(C_j) = \frac{1}{q} \quad \text{and} \quad P(\bar{\sigma} \mid C_j) = \frac{q(n - \bar{\sigma}(j))!}{n!(c - \bar{\sigma}(j))!}.$$

By (14), all orders of coalition formation are not equally probable; $P(\bar{\sigma} \mid C_j)$ decreases as $\bar{\sigma}(j)$ increases.

As usual, the extended Shapley value in the second model is obtained by substitution of (14) into the generalized Shapley value ((3) or (4)). For a quarreller, we have[4]

$$(15) \quad \phi_j^*[Q, v] = \sum_{\sigma \in \mathscr{S}(C_j)} \frac{(n - \bar{\sigma}(j))!}{n!(c - \bar{\sigma}(j))!} V_j(S_\sigma(j)) + v(j) \frac{q-1}{q} =$$

$$= \sum_{j \in S \subseteq C_j} (s-1)!(c-s)! \frac{(n-s)!}{n!(c-s)!} V_j(S) + v(j) \frac{q-1}{q} =$$

$$= \sum_{S \cap Q = j} \gamma(n, s) V_j(S) + \frac{q-1}{q} v(j).$$

As with (5) and (6), it is a simple exercise to describe (15) as j's expectation (cf. Lemma I).

Before discussing the extended Shapley value for a non-quarreller in the second model, we shall briefly develop some of the properties of $\phi_j^*[Q, v]$.

LEMMA III. $v(j) \leqslant \phi_j^*[Q, v] \leqslant \phi_j[v]$.

> Equality obtains in the first instance iff $V_j(S) = v(j)$ whenever $S \cap Q = j$.
> Equality obtains in the second instance iff $V_j(S) = v(j)$ whenever $S \cap Q \supset j$.

Proof. From Lemma I and (15) the first inequality follows, and the conditions necessary and sufficient for equality to obtain. From (2) and (15),

$$\phi_j[v] - \phi_j^*[Q, v] = \sum_{S \cap Q \supset j} \gamma(n, s) V_j(S) - \frac{q-1}{q} v(j).$$

By (13), the remainder of the lemma is now obvious.

Lemma III states that, in the second model of play, a quarreller cannot gain from the quarrel. As we have noted, this statement is false for the first model of play. Returning to the example above, it is clear that

$$\phi_1^*[Q, v] = \frac{c}{6};$$

$$\phi_1[v] - \phi_1^*[Q, v] = \frac{a + 2(1-b)}{6} \geqslant 0.$$

The necessary and sufficient conditions for equality, given in Lemma III, can also be verified in this case.

Now we obtain the extended Shapley value for a non-quarreller by substitution of (14) into (4). This yields

$$\phi_i^*[Q, v] = \sum_{j \in Q} \sum_{\sigma \in \mathscr{S}(C_j)} \frac{(n - \bar{\sigma}(j))!}{n!(c - \bar{\sigma}(j))!} V_i(S_\sigma(i)).$$

We evaluate the double summation by splitting it into two parts. First, let j, S, and h be fixed so that $j \in Q$, $i \in S \subseteq F$, and $s+1 \leqslant h \leqslant c$. Then there are exactly $(s-1)!\,(c-s-1)!$ choices of $\sigma \in \mathscr{S}\{C_j\}$ which satisfy $S_\sigma(i) = S$ and $\bar{\sigma}(j) = h$. Therefore

$$\sum_{\substack{j \in Q \\ \bar{\sigma}(i) < \bar{\sigma}(j)}} \sum_{\sigma \in \mathscr{S}(C_j)} \frac{(n - \bar{\sigma}(j))!}{n!(c - \bar{\sigma}(j))!} V_i(S_\sigma(i)) =$$

$$= q \sum_{i \in S \subseteq F} (s-1)!\,(c-s-1)! \sum_{h=s+1}^{c} \frac{(n-h)!}{n!(c-h)!} V_i(S) =$$

$$= q \sum_{i \in S \subseteq F} \frac{(s-1)!\,(c-s-1)!}{n!} V_i(S) \sum_{r=0}^{c-s-1} \frac{(n-c+r)!}{r!} =$$

$$= q \sum_{i \in S \subseteq F} \frac{\gamma(n,s)}{q} V_i(S), \quad \text{by (12)}$$

$$= \sum_{S \cap Q = \phi} \gamma(n,s) V_i(S).$$

Now fix j, S, and h satisfying $j \in Q$, $i, j \in S \subseteq C_j$, and $1 \leqslant h \leqslant s-1$. Then there are exactly $(s-2)!\,(c-s)!$ choices of $\sigma \in \mathscr{S}\{C_j\}$ which satisfy $S_\sigma(i) = S$ and $\bar{\sigma}(j) = h$. Therefore

$$\sum_{\substack{j \in Q \\ \bar{\sigma}(i) > \bar{\sigma}(j)}} \sum_{\sigma \in \mathscr{S}(C_j)} \frac{(n - \bar{\sigma}(j))!}{n!(c - \bar{\sigma}(j))!} V_i(S_\sigma(i)) =$$

$$= \sum_{j \in Q} \sum_{i, j \in S \subseteq C_j} \frac{(s-2)!\,(c-s)!}{n!} V_i(S) \sum_{h=1}^{s-1} \frac{(n-h)!}{(c-h)!} =$$

$$= \sum_{j \in Q} \sum_{i, j \in S \subseteq C_j} \frac{(s-2)!\,(c-s)!}{n!} V_i(S) \frac{1}{q} \left[\frac{n!}{(c-s)!} - \frac{(n-s-1)!}{(c-s)!} \right], \quad \text{by (12)}$$

$$= \sum_{|S \cap Q| = 1} \frac{\gamma(c-1, s-1) - \gamma(n, s-1)}{q} V_i(S).$$

We have shown that

$$\phi_i^*[Q, v] = \sum_{S \cap Q = \phi} \gamma(n, s) V_i(S) +$$
$$+ \sum_{|S \cap Q| = 1} \frac{\gamma(c-1, s-1) - \gamma(n, s-1)}{q} V_i(S).$$

We make some simple remarks which pertain to the interpretation of (16) as i's expectation. Since

$$\gamma(m+1, s) = \frac{m-s+1}{m+1} \gamma(m, s) < \gamma(m, s),$$

it follows that

$$\frac{\gamma(c-1, s-1) - \gamma(n, s-1)}{q} \geq 0.$$

The identities

(17) $$\sum_{\substack{S \cap Q = \phi \\ i \in S}} \gamma(n, s) = \sum_{\substack{|S \cap Q| = 1 \\ i \in S}} \frac{\gamma(n, s-1)}{q} = \frac{1}{q+1}$$

are consequences of (12), and can be derived analogously to Lemma I. Since the sum of the coefficients (of $V_i(S)$, when $i \in S$) in (16) is unity, it follows that

(18) $$\sum_{\substack{|S \cap Q| = 1 \\ i \in S}} \frac{\gamma(c-1, s-1)}{q} = 1.$$

Again, (18) is easy to prove directly. Note that, in contrast to (5), (6), and (15), it is difficult to check directly that (16) represents i's expectation. In the example above, it is clear that

$$\phi_3^*[Q, v] = \frac{b+c}{3}.$$

Therefore $\phi_3^*[Q, v] > \phi_3[v]$ iff

$$b + c > 1(1-a).$$

We can say, roughly, that 3 gains from the quarrel of 1 and 2 when 1 and 2 are very strong players. (Exactly the same statement was made within the first model of play.) We now attempt a similar analysis in the general case. From (2) and (16), we have that $\phi_i^*[Q, v] > \phi_i[v]$ iff

(19) $$\sum_{|S \cap Q|=1} \eta(n, q, s) V_i(S) - \sum_{|S \cap Q| \geq 2} \gamma(n, s) V_i(S) > 0,$$

where

$$\eta(n, q, s) = \frac{\gamma(c-1, s-1) - \gamma(n, s-1)}{q} - \gamma(n, s).$$

$\bigg($Clearly the sum of the coefficients (of $V_i(S)$, when $i \in S$) in (19) is zero. It can be shown, using (12), (17), and (18), that

$$\sum_{\substack{|S \cap Q|=1 \\ i \in S}} \eta(n, q, s) = \sum_{\substack{|S \cap Q| \geq 2 \\ i \in S}} \gamma(n, s) = \frac{q-1}{q+1}.\bigg)$$

LEMMA IV. $\eta(n, q, s) > 0$.

Proof. It is easy to verify that

$$\gamma(m, s) = \frac{m+1}{m-s+1} \gamma(m+1, s) = \frac{m-s}{s} \gamma(m, s+1).$$

Repeated application of these results shows that

$$\eta(n, q, s) = \frac{\gamma(n, s) \delta(n, q, s)}{q(s-1)},$$

where

$$\delta(n, q, s) = \frac{n!(n-q-s+1)!}{(n-q)!(n-s)!} - n - (q-1)(s-1).$$

That $\delta(n, q, 2) > 0$ can be checked easily. Suppose that $3 \leq s \leq n-q+1$ and that $\delta(n, q, s-1) > 0$. Then $\delta(n, q, s) > 0$, since

$$\frac{n!(n-q-s+1)!}{(n-q)!(n-s)!} = \frac{n!(n-q-s+2)!}{(n-q)!(n-s+1)!} \frac{(n-s+1)}{(n-q-s+2)}$$
$$> [n + (q-1)(s-2)] \frac{(n-s+1)}{(n-q-s+2)}$$
$$> n + (q-1)(s-1).$$

This proves the lemma.

Lemma IV and (19) justify the statement that, roughly speaking, i gains from the quarrel of Q when, for $i \in S$. $V_i(S)$ tends to assume large values when $|S \cap Q| = 1$, relative to its values when $|S \cap Q| \geq 2$. It is interesting to compare this statement with the results for the example, and the analogous statement for the first model of play.

We conclude with a brief comparison of the two models of play, in terms of the extended Shapley values derived from them. For convenience all of our Shapley values are repeated here. When there is no quarrelling,

$$(2) \qquad \phi_k[v] = \sum_S \gamma(n, s) V_k(S).$$

In the first model of play, when quarrelling occurs,

$$(5) \qquad \phi_j[Q, v] = \sum_{S \cap Q = j} \frac{\gamma(c, s)}{q} V_j(S) + \frac{q-1}{q} v(j);$$

$$(6) \qquad \phi_i[Q, v] = \sum_{S \cap Q = \phi} \gamma(c, s) V_i(S) + \sum_{|S \cap Q| = 1} \frac{\gamma(c, s)}{q} V_i(S).$$

In the second model of play, when quarrelling occurs,

$$(15) \qquad \phi_j^*[Q, v] = \sum_{S \cap Q = j} \gamma(n, s) V_j(S) + \frac{q-1}{q} v(j);$$

$$(16) \qquad \phi_i^*[Q, v] = \sum_{S \cap Q = \phi} \gamma(n, s) V_i(S) +$$

$$+ \sum_{|S \cap Q| = 1} \frac{\gamma(c-1, s-1) - \gamma(n, s-1)}{q} V_i(S).$$

First, we compare the two extended Shapley values for a quarreller. From (5) and (15),

$$\phi_j^*[Q, v] - \phi_j[Q, v] = \sum_{S \cap Q = j} \left[\gamma(n, s) - \frac{\gamma(c, s)}{q} \right] V_j(S).$$

Clearly, the value of $V_j(S)$ affects $\phi_j^* - \phi_j$ iff $S \cap Q = j$. In this case, it follows from Lemma II that an increase in $V_j(S)$ increases $\phi_j^* - \phi_j$ if s is small, and decreases $\phi_j^* - \phi_j$, if s is large. Equivalently, if $S \cap Q = j$, the probability that j joins $S - j$ is greater in the second model than in the first when s is small, and less when s is large.

Now we compare the two extended Shapley values for a non-quarreller. From (6) and (16),

$$\phi_i[Q, v] - \phi_i^*[Q, v] = \sum_{S \cap Q = \phi} [\gamma(c, s) - \gamma(n, s)] V_i(S) -$$
$$- \frac{1}{q} \sum_{|S \cap Q| = 1} [\gamma(c-1, s-1) -$$
$$- \gamma(c, s) - \gamma(n, s-1)] V_i(S) =$$
$$= \sum_{S \cap Q = \phi} [\gamma(c, s) -$$
$$- \gamma(n, s)] \left[V_i(S) - \frac{1}{q} \sum_{j \in Q} V_i(S \cap j) \right]$$

where we have used the easy identity

$$\gamma(c-1, s-1) = \gamma(c, s) + \gamma(c, s-1).$$

Clearly, $\phi_i - \phi_i^*$ depends on $V_i(S)$ iff $i \in S$ and $|S \cap Q| \leq 1$. If $i \in S$ and $S \cap Q = \emptyset$, the expression

$$V_i(S) - \frac{1}{q} \sum_{j \in Q} V_i(S \cap j)$$

is a measure of whether, on the average, the presence of one member of Q increases or decreases i's value to S. Of course, $\phi_i - \phi_i^*$ is a positive linear combination of all such expressions. We conclude that, if $i \in S$ and $|S \cap Q| \leq 1$, the probability that i joins $S - i$ is greater in the first model than in the second when $S \cap Q = \phi$, and less when $|S \cap Q| = 1$.

University of Toronto

NOTES

[1] Our derivation of the Shapley value (and the generalized Shapley value) is suggested by Shapley's observations in Chapter 6 of [1], p. 316.
[2] [1], p. 316.
[3] As the use of 'P' in our notation suggests, it is the probability P which is essential to our generalized Shapley value, (1), and not the model of play and the distributions of arrival time from which P is obtained. We might equally well have chosen another model in which (1) is player k's expectation, and where P would arise in an entirely different way. One such alternative is to make $P(\sigma)$ depend on some measure of the credibility of the first player in σ as a leader, combined with some measure of the appeal to the first player of the second player as a lieutenant, etc. From a mathematical

point of view, all of our models of play are merely convenient illustrations, through which the Shapley value and its extensions can be viewed as expectations. Only through P do they appear in the formalism.

[4] We have inserted an asterisk into our notation to distinguish the extended Shapley values derived from the second model of play from those derived from the first.

BIBLIOGRAPHY

[1] Shapley, L. S., 'A Value for n-Person Games', in *Contributions to the Theory of Games*, Vol. II (ed. by H. W. Kuhn and A. E. Tucker), Princeton University Press, Princeton, N.J., 1953, pp. 307–317.

JAMES D. LAING AND RICHARD J. MORRISON

COALITIONS AND PAYOFFS IN THREE-PERSON SUPERGAMES UNDER MULTIPLE-TRIAL AGREEMENTS

ABSTRACT. We outline here two formal models that predict coalitions and payoffs in sequential three-person games. The theories share the same structure, differing only in the planning horizon that we assume players use in calculating their strategies. Proceeding from *a priori* assumptions concerning the players' preference orderings over the various possible coalition outcomes and heuristic rules-of-thumb the players use in calculating their strategies, and assumptions about the nature of the bargaining process among the three players, the models predict both the probability that each coalition forms and the division of payoffs between coalition partners.

We analyze data from a laboratory study of coalition behavior in eleven male triads from the perspective of the theories. We summarize tests of the major predictions of the models with coalition outcomes dictated under the terms of agreements formed only for a single trial; these data offer strong and persistent support for the models. Then, turning to the primary focus of this paper, we generate successful predictions about coalitions and payoffs in the initiation, maintenance and disruption of long-term agreements to ally.

In this paper we develop theory predicting coalition outcomes in a sequential three-person laboratory game and report results from a pilot study of eleven male triads.[1] On each trial in the sequential game, a two-person coalition forms and divides the coalition's point value between the two partners. Each player knows that when the sequence of trials ends (without warning at a randomly chosen time), his monetary payoff for playing the game will be determined solely by the rank of his total point score accumulated across the sequence relative to the total scores of the other two players. Thus, the incentives in the laboratory situation encourage the players to seek maximum final rank. However, as is typical of sequential games played in natural or laboratory settings, our game is sufficiently complex that neither players nor analysts are able to represent the strategic situation as a whole in a form amenable to game-theoretic solution. Consequently, they are unable to develop strategies for playing each trial in the sequence that are demonstrably optimal with respect to maximizing final rank. Our models assume that players form a simplified representation of this complex situation, adopting as a surrogate objective the maximization of rank in the short run, hoping that

Anatol Rapoport (ed.), Game Theory as a Theory of Conflict Resolution, 207–234. *All Rights Reserved.*
Copyright © 1974 *by D. Reidel Publishing Company, Dordrecht-Holland*

this will lead to the maximization of final rank. The two models we employ in this paper use the same basic argument, although they differ in the planning horizon they assume players use. Proceeding from *a priori* assumptions about each player's decision calculus and the nature of the bargaining process among the three players, the models predict which coalitions form and how the partners will divide the points won by their coalition.

Two types of coalition agreements occur in the pilot study data. In a *single-trial agreement*, players agree to ally on the current trial, but make no commitments about future trials. In a *multiple-trial (n-trial) agreement*, players agree to form a coalition and to continue their alliance for at least one future trial. In an earlier paper (in press) we developed and tested two models that successfully predict pilot study outcomes formed under single-trial agreements. In the first portion of this paper we summarize the models and review the major results.[2] In most of the pilot study triads, although players struck single-trial agreements early in the sequential game, they eventually formed multiple-trial agreements. Such long-term agreements are important sources of coalition stability in the pilot study situation and, we think, in continuing social, political or economic relationships in general. The second main portion and the primary emphasis of this paper is devoted to the prediction and analysis of coalition outcomes dictated under the terms of multiple-trial agreements.

In the research described in this paper, we adopt rank maximization as a fundamental preference assumption. Given the way monetary prizes were awarded in the pilot study, we assert as an untested axiom that each player in this situation is motivated to maximize final rank. Thus, from one perspective, our research focuses on coalition behavior by rank-striving players. We believe that rank-striving is important in many natural and laboratory settings, even if rank-based motivations are not, in essence, 'imposed' by the demand characteristics of the situation. Consistent with Festinger's (1954) theory of social comparison processes, individuals and organizations frequently evaluate their achievement relative to the achievement of others. For example, if we are to take advertising statements seriously, Avis wants to be 'Number One'. A similar concern with rank is apparent in Kaplan's (1957) dictum for maintaining a 'balance of power' in international relations: "Act to oppose any coalition or single actor which tends to assume a position of predominance with respect to the rest of the system."

Moreover, rank-striving behavior is evident in a variety of laboratory studies of coalition behavior (Cf. Hoffman *et al.* (1954), Vinacke (1959, 1962), and Emerson (1964)). Thus, whether rank-striving behavior emerges 'naturally' or is 'imposed' by the demand characteristics of the situation, the research reported in this paper is motivated by the belief that research should be conducted to determine the effects of rank-based motivations on patterns of cooperation and conflict in continuing social relationships.

In outline, the argument of the paper proceeds as follows. Section 1 describes the two models. Section 2 summarizes the laboratory procedures used in the study. Section 3 reviews the major tests of the models' predictions using coalition outcomes dictated under single-trial agreements. Sections 4 and 5 then use the perspective of the theory to analyze coalition formation and payoff allocation, respectively, under multiple-trial agreements.

1. Two Models

This section briefly outlines the major features of the models we employ here.[3]

1.1. *Myopic Model*

Let Γ denote the set of essential three-person games described by the pseudo-characteristic function v:[4]

$$v(\emptyset) = v(I) = v(J) = v(K) = v(I, J, K) = 0, \text{ and}$$
$$v(I, J) = V_1, v(I, K) = V_2, v(J, K) = V_3$$

for all distinct players I, J, and K. V_1, V_2 and V_3 are positive even integers expressed in 'points'. Each coalition that forms exhausts all the points available to it, with payoffs to players constrained to be non-negative integers. The domain of the model is delimited to games in Γ occurring in sequences of one or more temporally disjoint games played by the same three players who are uncertain as to when the sequence will end.

We delimit the domain of the model to games in which players evaluate their performance relative to the success of the other players:

Let $\sigma_{iT} = \sum_{t=i}^{T} X_{it}$ denote the ith player's accumulated score through any trial T. We compute (in the conventional manner) r_{iT}, the *rank* of the

ith player's accumulated score in relation to the other players' scores. Then, letting the *interval position* of the ith player be denoted as $d_{iT} = \sum_j (\sigma_{iT} - \sigma_{jT})$ for $i \neq j$, $j \in \{1, 2, 3\}$, define the position of the ith player at the end of any trial as the two element vector (r_{iT}, d_{iT}). The model assumes that each player, say i, evaluates any outcome of a given trial T in the sequence according to the following lexicographic preference function which attributes primary importance to r_{iT}, his rank at the end of the trial:

Assumption 1.1. If any two positions differ in the rank they afford a player, he prefers the position with the higher rank.

Assumption 1.2. If any two positions afford a player the same rank but different interval positions, he prefers the position with the higher interval position.

Assumption 1.3. If two positions afford a player the same rank and interval position, they represent the same position, and he is indifferent between the outcomes.

Given these preferences and a game in Γ, the model assumes further that each player acts as if he employs bargaining heuristics that impose the following limits on the payoff he is willing, *a priori*, to grant his partner:[5]

Assumption 2.1. A player will not agree to a payoff vector which causes his coalition partner to overtake or pass him in rank.

Assumption 2.2. A coalition member will not agree to a payoff vector which not only fails to give him an improved rank but also causes him to lose interval position.

The model postulates that each player not only obeys these assumptions but also acts as if he expects the other players to obey them as well. Note that adherence to {2.1, 2.2} is sufficient to ensure that a coalition member attains a position at the end of the trial at least as preferred as his position at the beginning of the trial. The model also postulates:

Assumption 2.3. If a coalition member holds a rank-based preference for his partner, he will not accept a payoff smaller than that necessary to achieve the rank upon which this preference is based. ('Rank-based preferences' are defined below.)

Given the current position of each player, the function v describing the game to be played on a given trial, and the assumed bargaining heuristics that establish the lower bound for the minimum payoffs that we assume players deem acceptable, the myopic model solves the joint 'optimization' problem by the following sequence of operations. For each player, I, it calculates the initial upper bound on his payoff from a given coalition by identifying the maximum payoff, X_{ij}^+, that he can obtain from coalition (I, J) if his prospective partner, J, obeys the bargaining heuristics of Assumptions 2.1 and 2.2. The model identifies the initial aspirations of each member, I, of a prospective coalition (I, J), in terms of the rank afforded him were he to receive X_{ij}^+, and his partner, J, to receive the remainder $(v(I, J) - X_{ij}^+)$.

To check for joint feasibility of these initial aspirations, the model calculates the *minimum* payoff for each player in each of his coalitions $\{X_{ij}^-\}$ that enables him to attain the same rank as that associated with the maximum payoffs computed above, i.e., the X_{ij}^+.[6] If both partners in a potential coalition can attain at least these minimum payoffs without exceeding the point-value of their coalitions then neither need revise his rank aspirations for that coalition. Were the initial rank aspirations not mutually feasible, then at least one player would have to concede by accepting a lower rank than that associated with his initial aspirations. However, the following result guarantees that neither player need concede to a lower rank: *If Assumption 2.1 is not violated, then the initial rank aspirations held by prospective coalition partners are mutually feasible.* Therefore, each player I maintains his aspiration to achieve the rank (r_{ij}^+) associated with the payoffs in the closed interval (X_{ij}^-, X_{ij}^+) and the coalition (I, J).

If player I prefers the rank he expects from coalition (I, J) to that from (I, K), that is, $r_{ij}^+ > r_{ik}^+$, then I prefers J as coalition partner and chooses J as partner with 'Attraction' probability $A_{ij} = 1 - \varepsilon$, choosing K with probability $A_{ik} = \varepsilon$, where $0 < \varepsilon < \frac{1}{2}$. If a player has no rank preference, i.e. $r_{ij}^+ = r_{ik}^+$, then he chooses J and K with probabilities proportional to the likelihood that they choose him, i.e. $A_{ij}/A_{ik} = A_{ji}/A_{ki}$.[7] Using the concept of reciprocated choice as the necessary and sufficient condition for coalition formation and these choice probabilities, we compute the probability that coalition (I, J) forms:

$$P^*(I, J) = \frac{A_{ij}A_{ji}}{A_{ij}A_{ji} + A_{ik}A_{ki} + A_{jk}A_{kj}}$$

for distinct $i, j, k \in \{1, 2, 3\}$, the set of players.

We next develop the predictions as to how partners will divide the payoffs. First of all, the model predicts that the payoff from coalition (I, J) to player I lies in the closed interval $[X_{ij}^-, X_{ij}^+]$, called the *negotiation range* for the (I, J) coalition. Note that if coalition (I, J) forms and the partners agree to a division in the negotiation range, then both partners realize their rank aspirations (r_{ij}^+, r_{ji}^+) for the coalition. (Note further that any division within the negotiation range satisfies Assumptions 2.1, 2.2 and 2.3 for both partners. Any division outside the negotiation range violates at least one of these assumptions.) This implies that each partner's preferences for outcomes in the negotiation range are monotone increasing with his payoff in points; that is, partner I's preferences increase as his payoff increases from X_{ij}^-. Since $X_{ji} = v(I, J) - X_{ij}$, the two partners have strictly opposing preferences for outcomes within the negotiation range for their coalition.

To predict payoffs within the negotiation range, the model examines the attraction structure for asymmetries that give one player a bargaining advantage over his partner. Articulating a concept from Emerson's (1962) model of dependence relationships,[8] let the *dependence* of player I on player J as a coalition partner be defined as $D_{ij} = A_{ij}/A_{ki}$ for all distinct I, J, and K. Player I has a *bargaining advantage* over player J if and only if I is less dependent on J than is J on I, i.e. $D_{ij} < D_{ji}$. The theory states that the conflict between coalition partners in selecting a payoff division within the negotiation range tends to be resolved in favor of the player with the bargaining advantage:

> A coalition member who enjoys a bargaining advantage over his partner has the ability to demand outcomes up to his most preferred payoff within the negotiation range, and will tend to do so. If no bargaining advantage is identified within the coalition, then the partners will tend to agree to that payoff payoff configuration nearest the mean of the negotiation range.

Thus, taking as input the function v of a game in Γ to be played on any trial in the sequence of games, the players' positions at the beginning of the trial, and the value for the parameter, ε, the myopic model states the probability that each coalition forms and predicts the payoff division for each of the three coalitions.

These arguments complete the description of the *myopic model*, so named because we assume the players are short-sighted, looking only to the end of the current trial in their calculations. The second model we employ here, the *hyperopic model*, is identical to the myopic model in all but one respect: it employs a different game 'value' to ascertain rank-based preferences in a (perhaps excessively) farsighted planning horizon.

1.2. *Hyperopic Model*

Suppose that, because of their limited computational ability during coalition negotiations, players use simple extrapolation to forecast future outcomes. They assume that whatever coalition forms on this trial will continue to form on future trials. By this assumption we might regard the point 'value' of a coalition in the current game as being progressively increased as we add future games into our computations. That is, if we assume that the same coalition forms for the next n trials, then the apparent 'value' in points of the (I, J) coalition in the current game is the sum of the $v(I, J)$ values of games to be played for the next n trials. As the apparent point value of the coalitions in the current game (and the associated planning horizon) increases, eventually these values become sufficiently large that any effect of the distances between players' accumulated scores is overcome. Consequently no additional change occurs in the players' attractions for each other as coalition partners if the apparent point values continue to increase. We refer to this final network of choice probabilities as the *limiting attraction structure*. The hyperopic model assumes that players make their choices of coalition partners on the current trial according to the choice probabilities of the limiting attraction structure. (These probabilities are determined on the basis of the ranks held by the players at the beginning of the trial and a straightforward application of Assumption 2.1.) There are only four types of possible distributions of ranks among three players, and each distribution has a unique limiting attraction structure. Thus, the hyperopic model identifies a distinct (limiting) attraction structure for each distinct distribution of ranks, as shown in Figure 1. Once we find the appropriate attraction structure, we proceed as in the myopic model with respect to the attribution of dependence and bargaining advantage, and the prediction about what coalition forms and how partners divide the (nominal) value of the current game.

Distribution of Ranks at Beginning of Trial	Limiting Attraction Structure
(1) $\sigma_A = \sigma_B = \sigma_C$	A with arrows: from A to B (0.5), from A to C (0.5); from B to A (0.5), from B to C (0.5); from C to A (0.5), from C to B (0.5)
(2) $\sigma_A = \sigma_B > \sigma_C$	From A: (ε) to B, $(1-\varepsilon)$ to C; from B: (ε) to A, $(1-\varepsilon)$ to C; from C: (0.5) to A, (0.5) to B
(3) $\sigma_A > \sigma_B > \sigma_C$	From A: $\dfrac{\varepsilon(2-\varepsilon)}{1+\varepsilon}$ to B, $\dfrac{1-\varepsilon+\varepsilon^2}{1+\varepsilon}$ to C; from B: (ε) to A, $(1-\varepsilon)$ to C; from C: $\dfrac{1-\varepsilon^2}{2-\varepsilon}$ to B, $\dfrac{1-\varepsilon+\varepsilon^2}{2-\varepsilon}$ to A
(4) $\sigma_A > \sigma_B = \sigma_C$	From A: (0.5) to B, (0.5) to C; from B: (ε) to A, $(1-\varepsilon)$ to C; from C: (ε) to A, $(1-\varepsilon)$ to B

Fig. 1. Probabilities of choosing coalition partners according to *hyperopic* model, under four conditions.

2. Pilot Study Laboratory Procedures

Subjects in the pilot study were 33 male masters students in a quantitatively oriented business school program. We constituted the eleven triads randomly from the population of volunteers. Each triad played one

COALITIONS AND PAYOFFS IN THREE-PERSON GAMES 215

sequence of games in Γ. The value of each game ($V = V_1 = V_2 = V_3$) was expressed in 'points', with single points indivisible. The subjects knew that the values of the various games in the sequence would differ, with the game values chosen from a table of random numbers using the following probabilities: $p(V=100) = \frac{1}{2}$, $P(V=300) = \frac{1}{3}$, and $p(V=500) = \frac{1}{6}$. The play involves face-to-face bargaining within the triad to achieve an agreement specifying which two players form a coalition and how they divide the points between them. After each trial, the investigator updates the current accumulated scores on the blackboard in full view of the subjects, erasing the previous scores so that no historical record of previous play is displayed. Subjects know that the sequence of games ends without warning at a randomly chosen time, with each player's share of the total $15 in prizes being a positive linear function of his rank in the final distribution of accumulated scores, ranging from $0 for sole posession of last place to $10 for first place.

3. COALITIONS AND PAYOFFS UNDER SINGLE TRIAL AGREEMENTS

The myopic and hyperopic theories explain and predict supergame effects on coalition behavior during each trial in a sequential game. Generally, since the theories attempt to account for the systematic interdependence of trials, we think it legitimate to treat each trial within the sequence as an 'independent event', except for the interdependence noted by the theory under investigation.

This assumption, however, does not appear tenable if the outcomes of several trials are dictated under the terms of a single agreement. On the basis of the investigator's notes taken during the bargaining sessions of the pilot study and a careful review of the audio tapes of these sessions, we judge for each trial outcome whether or not that outcome is dictated by an agreement whose terms include the outcome of more than one trial.[9] We refer to these as *n-trial agreements*. (We made these coding decisions *prior* to data analysis.) Since *n*-trial agreements are strong exogenous sources of interdependence if we use the outcome of a single trial as the unit of analysis, we thus admit only those 76 trial outcomes which are not dictated by an *n*-trial agreement for testing hypotheses derived from the theories.[10] On the other hand, long term agreements to ally are an important source of coalition stability in the laboratory as well as in natural arenas. In the following sections we first test hypotheses derived from the theory with the 76 pilot study outcomes dictated by an agreement pertaining only to the current trial. Then we use the theory as a perspective for analyzing coalition formation and payoff divisions dictated by long-term agreements.

3.1. *Coalition Formation under Single-Trial Agreements*

Our first task is to estimate the parameter ε to test the hypothesis that the $0 < \varepsilon < \frac{1}{2}$ and to measure the goodness-of-fit the two theories achieve in the pilot study data. To estimate the parameter, we choose that value of ε that maximizes the probability of the exact distribution of coalitions observed during the pilot study. Letting $P^*(a, c, \varepsilon)$ represent the probability stated by the theory as a function of ε that coalition c forms in attraction structure a, select ε to maximize the likelihood function

$$L = \prod_a \left[K_a \prod_c P^*(a, c, \varepsilon)^{N(a, c)} \right],$$

where $N(a, c)$ denotes the number of times coalition c forms under attraction structure a during the 76 pilot study trials, and $K_a = = [\sum_c N(a, c)]! / \prod_c [N(a, c)!]$. The resulting maximum value of L is the natural (maximum likelihood) goodness-of-fit measure for the theory under investigation. To compare the theory to the null hypothesis that each two-person coalition is equally likely, we evaluate the ratio L/L_0, where L_0 is the value for L when $P^*(a, c, \varepsilon_0 = 0.5) = \frac{1}{3}$.

Table I shows the pilot study results for the two models. For each model the data support unambiguously the hypothesis that $0 < \varepsilon < \frac{1}{2}$. Moreover, the ratio L/L_0 shows that the distribution of observed coalitions is 49(117) times more likely under the myopic (hyperopic) model than under the null hypothesis.[11]

TABLE I

Maximum likelihood results for two models in the pilot study data ($N = 76$)

	Myopic model	Hyperopic model
ε	0.35	0.33
$L/L_0 =$	49	117

Although the pilot study yields too few cases to permit analysis for each attraction structure separately, one attraction structure occurs frequently enough that we can use it for illustrative purposes. The data we show in Table II for this structure display good correspondence to the predicted distributions.

TABLE II

Pilot study results for two models when, solely on the basis of rank considerations, B is identified as preferring C, while A and C are identified as being indifferent between prospective coalition partners, showing values of the residual error (e)

	Coalition			
	(A, B)	(A, C)	(B, C)	Total
Myopic model:				
Observed, $P(I, J) =$	0.143	0.357	0.500	1.000
Predicted, $P^*(I, J) =$	0.196	0.351	0.453	1.000
Error, $e = P(I, J) - P^*(I, J) =$	−0.053	0.006	0.047	Mean ABS. error = 0.035
$(N) =$	(6)	(15)	(21)	(42)
Hyperopic model:				
Observed, $P(I, J) =$	0.155	0.345	0.500	1.000
Predicted, $P^*(I, J) =$	0.178	0.356	0.466	1.000
Error, $e = P(I, J) - P^*(I, J) =$	−0.023	−0.011	0.034	Mean ABS. error = 0.023
$(N) =$	(9)	(20)	(29)	(58)

Note: For all cases in Table II, $\sigma_A > \sigma_B > \sigma_C$.

The results shown in Tables I and II depend on the value of ε estimated for the pilot study data. However, given the *a priori* assumption that $0 < \varepsilon < \frac{1}{2}$ the models make ordinal predictions concerning the probability that each coalition forms that do not depend on the particular value assumed by ε within the interval $(0, \frac{1}{2})$. When reduced to ordinal form, the hyperopic model makes predictions concerning coalition formation that are nearly equivalent to Emerson's (1964) predictions. [For example, if $\sigma_A > \sigma_B > \sigma_C$, the hyperopic model predicts $P^*(B, C) > P^*(A, C) > > P^*(A, B) > 0$), while Emerson (1964), p. 295) predicts "Not AB; BC more than AC."]

The results shown in Table III offer strong support for these ordinal coalition predictions.

TABLE III

Tests of ordinal predictions concerning coalition formation by two models using pilot study data

Proportion of times indicated coalition forms when model predicts that coalition to be the single		
	Myopic model	Hyperopic model
most likely:	0.481 (25/52)[a]	0.500 (30/60)[b]
least likely:	0.193 (11/57)[a]	0.172 (11/64)[c]

Note: using the binomial test, these proportions differ from $P(I, J) = \frac{1}{3}$ in the predicted direction at the following levels of significance:

[a] $p < 0.02$,   [b] $p < 0.01$,   [c] $p < 0.005$

In summary, though we test only the most general coalition predictions of the two models with these pilot study data, both models enjoy strong and persistent support. We note in passing that the hyperopic model better accounts for these data, but too few cases are available to place the two models in direct competition with each other. We conclude that each model is successful in predicting coalition formation in the pilot study.

3.2. *Payoff Allocation under Single-Trial Agreements*

The most important payoff prediction of the models states that if the model identifies one partner as enjoying a bargaining advantage in the coalition, then he has the ability to obtain payoffs up to his most preferred outcome within the negotiation range, and tends to do so. To test this prediction, we define the index of Negotiation Success (*NS*) achieved by the *i*th player,

$$NS_i = \frac{X_{ij} - X_{ij}^-}{X_{ij}^+ - X_{ij}^-} \quad \text{for} \quad X_{ij}^+ > X_{ij}^-.$$

X_{ij} is the payoff *I* obtains in coalition (I, J). If a coalition member achieves his maximum for that coalition, his *NS* value is unity, while if he receives his minimum the *NS* value is zero. As a consequence of the way we define the negotiation range, if coalition (I, J) forms, then $NS_j = 1 - NS_i$ (indicating the partners' strictly opposing preferences in dividing the payoffs). If NS_i (or NS_j) is negative or greater than one, then the division of payoffs is 'excessive' and represents a violation of Assumption 2.1, 2.2 or 2.3 by one of the players.

Note that NS is undefined if the negotiation range consists of a single point; this occurs for 13 cases in the pilot study data. In each of these cases, both models predict an even split of the payoffs: seven resulted in even splits, 8 are within one point and eleven are within 5 points of a 50–50 division. Of the remaining 63 cases, only 22% result in even splits, and these divisions typically are consonant with the predictions.

TABLE IV

Payoff allocation to player with bargaining advantage under two models, pilot study data

Bargaining advantage identified by: Variables describing payoffs to advantaged players:	Myopic model	Hyperopic model
Average NS	0.859 [b]	0.820 [b c]
Median NS	0.85	0.83
Proportion $NS > 0.5$	0.80 [b]	0.76 [b]
Proportion $NS \geqslant 0.5$	0.82 [b]	0.79 [b]
(N)	(55)	(63)
Proportion of *uneven* splits in favor of advantaged player	0.79 (34/43) [b]	0.75 (36/48) [b]
Proportion of *excessive* payoffs in favor of advantaged player	0.94 (15/16) [b]	0.79 (15/19) [a]

Note: Average NS results tested by t-test against two null hypotheses: $NS = 0.5$, and NS if each coalition had split payoffs evenly. All proportions were tested against $P = \frac{1}{2}$ by binomial test. Significance levels thus obtained are:

[a] $p < 0.01$, [b] $p < 0.001$

[c] Average NS for hyperopic model omits one extreme case ($NS = 17.00$) to avoid misrepresenting results. This case omitted *only* when mean NS is being computed.

The first four rows of Table IV test the major payoff hypothesis of the models. These results offer strong support for this hypothesis: on average, the advantaged player achieves a payoff more than 80% toward his most preferred end of the negotiation range, and roughly 80% of the advantaged players obtain payoffs in their most preferred half of the range.

Row five of Table IV shows strong support for the bargaining advantage argument independent of the negotiation range. Too few cases are available in the pilot study data for definitive contrasts between the myopic and hyperopic models. It is interesting to note that Emerson's (1964) predictions concerning payoff division are equivalent to the hypothesis that uneven payoffs favor the advantaged player that the hyperopic model identifies.

Row six suggests that bargaining advantage is a good predictor of which player is likely to receive the benefit of payoffs outside the negotiation range. (Recall that these 'excessive' payoffs are violations of Assumptions 2.1, 2.2 or 2.3.) Although the pilot study provides too few cases in which the models disagree to permit reliable contrasts, on the basis of our analyses we speculate that there is an interaction effect of the bargaining advantages attributed by the two models. Players appear to act as if they consult both models for signals as to the appropriate division; the more these signals agree, the more predictable are the payoff allocations.

We take these results to offer strong support for both the myopic and hyperopic predictions as to what coalition forms and how payoffs are divided in those pilot study outcomes dictated under the terms of a single-trial agreement. [For more extensive analysis and discussion, see Laing and Morrison (in press).]

4. Coalition Formation and n-Trial Agreements

We argue in Section 2.2 that only coalition outcomes dictated under single-trial agreements should be used for testing hypotheses derived from the theory. However, in accounting for coalition behavior during the pilot study, we should not ignore agreements dictating the outcomes of two or more trials. Clearly, analogues to n-trial agreements ($n \geqslant 2$) are an important source of coalition stability, whether in natural or artificial settings. The significance of n-trial agreements in coalition behavior during the pilot study is unmistakable, since 62% (123/199) of the single trial outcomes during the pilot study are dictated under the terms of an n-trial agreement to ally. Altogether twenty-one coalitions form an n-trial agreement during the pilot study. These agreements are distributed fairly evenly across the eleven triads: the number of n-trial agreements per triad ranges from zero to four, the median number is two, and this is also the mode with fully six of the eleven triads in this category. In the modal case one of the members of the initial n-trial agreement 'double-crosses' his partner and a second n-trial agreement forms.

In the following sections we investigate when and which players initiate and defect from n-trial agreements to ally.

4.1. *Initiation of n-Trial Agreements*

Here we focus on *first installment outcomes*, defined as the first (single-trial) outcome in a series of trials in which the same two players ally under an n-trial agreement. Since the pilot study data provide only 21 first installment outcomes, our analysis of the initiation of n-trial agreements must be limited to overall characteristics, and is only suggestive.

First, the models' hypotheses concerning what coalition forms predict which coalitions initiate n-trial agreements. Using values for the parameter (ε) estimated from the 76 pilot study outcomes dictated under a single-trial agreement (see Table I), we can predict the probability distribution of the various coalitions in the first installment outcomes, and measure our prediction success by evaluating the likelihood ratios. By this approach, neither the myopic nor the hyperopic model enjoys much success. Both likelihood ratios are close to unity, indicating that these models do not make predictions substantially better than the null hypothesis that the three coalitions are equally likely.

To predict *when* players initiate n-trial agreements, we assume that players associate risk with the introduction of n-trial agreements, since such agreements increase the apparent value of the current game, hence increase the magnitude of one's losses if he is excluded from the agreement. If attraction structures of the myopic and hyperopic models agree, then there is no one whose bargaining situation changes by playing the trial 'hyperopically', and hence (we assume) no one is likely to be willing to incur the risks of introducing the possibility of an n-trial agreement. That is, we predict that *the agreement of the models' attraction structures tends to be a sufficient condition for preventing the initiation of n-trial agreements*. This statement may be taken to predict that the entry in the lower left-hand cell of Table V is substantially smaller than expected under the null hypothesis of statistical independence. As predicted, the observed frequency (6) in this cell is significantly ($p<0.01$) smaller than the number expected under the null hypothesis.[12]

4.2. *Defection from n-Trial Agreements*

In this section we focus upon the outcomes of *defection trials*, trials in

TABLE V

		Attraction structures of myopic and hyperopic models	
		Agree	Disagree
Agreement made in current trial dictates the outcome of	Current trial only	54	16
	n-trials	6	9

N.B. Entries in the second row represent the initiation of n-trial agreements. Outcomes are admitted to this table only if the outcome of the immediately proceeding trial is not dictated under the terms of an n-trial agreement.

which the coalition that formed under an n-trial agreement on the immediately preceeding trial fails to form. First, we investigate the extent to which our models predict successfully *which partner defects*, then test an hypothesis predicting *when* defections occur.

Twelve of the twenty-one n-trial agreements terminate before the end of the sequential game with a defection by one of the partners to the agreement. At the beginning of all the defection trials, the previously excluded player is in last place in the distribution of accumulated scores (hence is named 'C'). As a consequence, we must convert the theoretical probabilities of each coalition (using the ε values estimated earlier, Table I) to conditional probabilities given that (A, B) can not form on these defection trials. Eliminating two defection trials in which the names of the players are arbitrary because of ties in rank, only ten defection trials from the pilot study remain for our investigation. Compounding this difficulty, the models make predictions indistinguishable with so few cases from the null hypothesis that the two possible coalitions are equally likely. The theoretical probability that (A, C) forms given (A, B) cannot form is 0.43 according to the hyperopic model and 0.48 (through a weighted average of attraction structures) by the myopic model. (Obviously, the conditional probability of $(B. C)$ given not (A, B) for each model is computed by subtracting these numbers from unity.) Five of each of the two possible coalitions form on defection trials; using the binomial test, we cannot reject the models' predictions, although neither can we reject the null hypothesis.

To predict *when* defections occur, we state the *ad hoc* (but *a priori*)

TABLE VI

Defections as a monotone increasing function of trial value, for trials in which the previous outcome was dictated under an n-trial agreement

	Trial value, $V=$		
	500	300	100
Defection	0.192 (5)	0.130 (3)	0.062 (4)
No Defection	0.808 (21)	0.870 (20)	0.938 (60)
	1.000 (26)	1.000 (23)	1.000 (64)

Goodman and Kruskal's gamma $= +0.44$;
Point-biserial correlation 1 0.29, $p<0.01$.

hypothesis that the probability of defection on a trial is monotone increasing with the point-value of the game to be played on that trial. The data we show in Table VI support this hypothesis.

5. Payoff Allocation and n-Trial Agreements

In the following subsections we use the bargaining advantage arguments of the myopic and hyperopic models to predict and to explain the division of payoffs between partners in initiating, maintaining and defecting from n-trials agreements.

5.1. *Payoff Division and INITIATION of n-Trial Agreements*

Given the scarcity of initiation trials in the pilot study data, we must regard our results from this analysis as highly tentative. Further, on intuitive grounds, we expect that there is likely to be 'noise' associated with the initiation of n-trial agreements. That is, once players begin negotiations for an n-trial agreement, it becomes especially important to be included in any resulting coalition; this encourages players to place even less emphasis on payoff divisions than we might assume, at least until the initial agreement to ally is made. Players might accept an initially disadvantageous payoff division to assure admittance to a putatively stable coalition, hoping to renegotiate the agreement later to achieve a more favorable share of the coalition's proceeds. This is consistent with our observation that n-trial agreements in the pilot data typically do not include explicit formulae for the division of future payoffs. The reader will recall that, *a priori*, we expect players to obey Assumptions 2.1–2.3 except *under duress*.[13] Presumably the threat of exclusion from a long-term coalition may cause sufficient duress that a player might reluctantly accept less in a first installment outcome than the models assume. In general, the emergence of 'stable' payoff configurations presumedly is more likely the more complete and unrestrained is the bargaining process.

An interesting (if incidental) feature of the payoffs during the initiation of n-trial agreements is the extent to which partners tend to split the payoffs evenly. Even splits occur in ten of the twenty-one first installment outcomes. This tendency is stronger than that observed in the pilot study trials whose outcomes are dictated by single-trial agreements. Once more the models lend some assistance in accounting for these results: when the myopic (hyperopic) model predicts an even split, it occurs in 4 out of 6 (3 out of 3) cases. Moreover, each of the ten even splits observed in first installment outcomes falls within the negotiation range established by the models.

Table VII shows data that are directly relevant to the models' payoff predictions for the first installment outcomes. Again, too few cases are available to place the models in competition with each other. Although the results are not strong and are based on few observations, the data consistently lie in the direction that the models predict.

TABLE VII

First installment payoffs: Bargaining advantage and payoffs during trials which initiate an n-trial agreement between two players, where *NS* denotes the index of negotiation success, and 'excessive' payoffs lie outside the negotiation range

Variables describing payoffs to advantaged players:	Bargaining advantage identified by:	
	Myopic model	Hyperopic model
Average *NS*	0.60	0.57
Median *NS*	0.70	0.62
Proportion $NS > 0.5$	0.67 (n.s.)[a]	0.56 (n.s.)
Proportion $NS \geqslant 0.5$	0.80 ($p < 0.02$)	0.72 ($p < 0.05$)
(*N*)	(15)	(18)
Proportion of excessive payoffs in favor of the advantaged player	0.67 (4/6) (n.s.)	0.62 (5/8) (n.s.)
Proportion of uneven splits in favor of advantaged player	0.56 (5/9) (n.s.)	0.64 (7/11) (n.s.)

[a]Not significant. However, considering only cases having $NS \neq 0.5$, $p < 0.05$.

5.2. Payoffs DURING n-Trial Agreements

In our previous analyses, we take as our empirical unit of analysis the outcome of a single trial within the sequential game, assuming that the outcomes are otherwise independent events, except as accounted for by the models under investigation. This assumption is no longer tenable once we begin analysis of coalition processes during sequences of outcomes dictated under an n-trial agreement. In this section we employ two alternative units of analysis. First, we treat each series of outcomes dictated under an n-trial agreement as a single *composite game* whose game value equals the sum of all games within the series of trials played under an n-trial agreement. Using the players' positions at the beginning of the n-trial agreement and the value of the composite game, sufficient data are available to derive the predictions for each of the models. Second, we examine single trial outcomes contained in the composite games, reporting separate results for the two units of analysis.

The pilot study provides only 21 composite games, with game values ranging from 100 to 5000 points, an average value of approximately 1280 points, and a median of 900. Altogether, 123 single-trial outcomes are contained in these composite games. Given the scarcity of composite games in the pilot study data we again must regard our results as tentative.

Again, we digress briefly to consider equal divisions of payoff. Even splits occur in fully two thirds (82/123) of the single trial outcomes within composite games, although less than 40% (8/21) of the composite games result in exactly 50–50 divisions. However, our analyses suggest that any 'natural' tendency towards even splits during n-trial agreements is affected strongly by the extent to which such an outcome is consistent with our models. For example, all of these even splits fall within the negotiation range identified by the models. Moreover, in those cases for which the models state that only an even split is permissible $[X_{ij}^+ = X_{ij}^- = V/2]$, then 50–50 divisions occur without fail! (The proportions are 26/26 for single trial outcomes within composite games, and 2/2 for composite game outcomes.) Table VIII admits all n-trial agreement outcomes (not only those for which the negotiation 'range' consists only of a single point, $V/2$). The models enjoy marked success in these results (although we suffer from an inadequate number of composite games), and the models enjoy increasing success the more similar are their predictions. Thus, there appears to be an

TABLE VIII

Interaction effects of two models in predicting even splits during n-trial agreements, showing pilot study results for two alternative units of analysis

Unit of analysis:	Single trial outcome during n-trial agreement				Composite game[c]		
	Even	Approx. Even[a]	Uneven		Even	Approx. Even[a]	Uneven
			Even	Uneven			
Division of payoffs predicted by: Hyperopic model Myopic model							
Observed payoff:							
$= V/2$	28	11	26	17	2	1	5
$\sim V/2$ [b]	7	0	0	1	1	0	0
$\neq V/2$	0	1	15	17	0	1	10
Goodman and Kruskal's gamma		+0.49				+0.59	

[a] Predicted division uneven but within four payoff points of even split.
[b] Observed division uneven but within one payoff point of even split.
[c] The two models disagreed in their payoff predictions (as described) for only one of the composite games; this case has been omitted from this table to simplify the exposition.

Note that the observed proportion of even splits decreases as the models vary from agreement in predicting even splits to agreement in predicting uneven splits.

interaction effect, with players acting as if they consult both models for signals, thus behaving in increasingly predictable ways the more these signals agree.

Additional analyses of the even splits during n-trial agreements based on the index of negotiation success, NS, suggest that there is no essential difference between these 50–50 divisions and other payoff allocations when viewed from the perspective of the models, (Laing and Morrison, 1971). We interpret these results to indicate that there is no strong tendency towards even splits in the pilot study data, including n-trial agreement outcomes, that is not in harmony with our predictions.

This completes our digression to investigate even splits. We turn now to our primary purpose, a general analysis of the extent to which the models explain payoff allocations during n-trial agreements.

TABLE IX

Payoffs during n-trial agreements to players enjoying a bargaining advantage, showing results under two units of analysis

Unit of analysis:	Single trial outcome dictated by an n-trial agreement		Composite game
Bargaining advantage identified by:	Myopic model	Hyperopic model	Both models
Variables describing payoffs to advantaged players:			
Average NS	0.83	0.81	0.83
Median NS	0.80	0.54	0.92
Proportion $NS > 0.5$	0.76 ($p < 0.001$)	0.54 (n.s.)[a]	0.78 ($p < 0.02$)
Proportion $NS \geqslant 0.5$	0.87 ($p < 0.001$)	0.88 ($p < 0.001$)	0.83 ($p < 0.01$)
(N)	(55)	(96)	(18)
Proportion of *excessive* payoffs in favor of advantaged players	0.84 (16/19) ($p < 0.003$)	0.87 (26/30) ($p < 0.001$)	0.88 (7/8) ($p < 0.04$)
Proportion of *uneven* splits in favor of advantaged players	0.81 (21/26) ($p < 0.003$)	0.76 (31/41) ($p < 0.001$)	0.83 (10/12) ($p < 0.02$)

N.B. NS based results omit cases for which NS is undefined. This occurs for coalition (I, J) in the event $X^-_{ij} = X^+_{ij}$.

[a] Not significant. However, considering only cases having $NS \neq 0.5$, $p < 0.001$.

Table IX shows that payoffs observed during n-trial agreements display strong support for our predictions based on the concept of bargaining advantage. These results are essentially identical to those we report earlier in this paper for those pilot study outcomes dictated by single-trial agreements. The one exception is the marginal support for the hyperopic model shown in the second and third entries of the second column of Table IX, although the data lie in the predicted direction. Apparently this attenuation of the degree to which the pilot study subjects successfully exploit their bargaining advantage identified by the hyperopic model can be attributed to interaction effects, as shown in Table X. Once more we

TABLE X

Interaction effects on payoffs of bargaining advantages attributed by myopic and hyperopic models, where results are based on single trial outcomes dictated by an n-trial agreement as the unit of analysis

Player I's bargaining advantage according to:[a]			
Hyperopic model		Advantage	
Myopic model	Advantage	Neither advantage nor disadvantage	Disadvantage
Variables describing payoff to player I:			
Average NS	0.88	0.79	0.46[b]
Median NS	0.93	0.50	0.50
Proportion $NS > 0.5$	0.85 ($p < 0.001$)	0.24 (n.s.)	0.25 (n.s.)
Proportion $NS \geqslant 0.5$	0.89 ($p < 0.001$)	0.88 ($p < 0.001$)	0.75 (n.s.)
(N)	(47)	(41)	(8)
Proportion of *excessive* payoffs in favor of I	0.84 ($p < 0.01$)	0.91 ($p < 0.01$)	–
	(16/19)	(10/11)	(0/0)
Proportion of *uneven* splits in favor of I	0.83 ($p < 0.001$)	0.67 (n.s.)	0.50 (n.s.)
	(20/24)	(10/15)	(1/2)
Proportion of even splits	0.49	0.63	0.75
	(23/47)	(26/41)	(6/8)

[a] The pilot study data provide no cases during n-trial agreements for the other combinations of bargaining advantages attributed by the models; hence these columns have been omitted from the table.

[b] Since, given coalition (I, J), $NS_i = 1 - NS_j$, Average $NS_J = 1 - 0.46 = 0.54$ for player identified by hyperopic model is *dis*advantaged *and* by mopic model as advantaged. All other results in the third column are identical regardless of which partner's payoffs are analyzed.

observe that players are more likely *to exploit* any bargaining advantage they are said to possess if both models agree in identifying them as advantaged. Two interpretations of this persistent pilot study result seem promising. First we assume that players act as if they consult both models to determine their situations. If both models attribute a bargaining advantage to the same player, then the players receive consistent signals concerning the appropriate allocation within the coalition and are likely to behave accordingly. If only *one* of the models identifies an advantaged player, then the absence of a similar attribution by the other model attenuates the effect of the bargaining advantage. If the models make *opposite* attributions of advantage, then the players are receiving confused signals, resulting in a more random allocation of payoffs within the negotiation range.

A second interpretation of these interaction effects is that players are neither myopic nor hyperopic in their calculations, but rather use an intermediate planning horizon in assessing the bargaining situation. One method of incorporating this assumption into our theory requires the addition of a new parameter, ψ, where $0 \leq \psi \leq 1$. Let M_{ij} and H_{ij} denote A_{ij} as identified by the myopic and hyperopic models, respectively. Then define a new attraction matrix, $\{A_i^*\} = \{\psi M_{ij} + (1-\psi) H_{ij}\}$. We can use this new matrix of attractions to compute the probability that coalition (I, J) forms and to identify bargaining advantages and associated payoff predictions. Lack of data requires that in this paper we refrain from any attempt to employ ψ in our analysis.

In any event, the data from the pilot study outcomes provide consistently strong support for the payoff predictions of the models, regardless of whether or not those outcomes are dictated by an *n*-trial agreement. Next we investigate the extent to which the models enable us to account for pay-offs during trials in which one player *defects* from an *n*-trial agreement.

5.3. *Payoffs during DEFECTION Trials*

Although we suffer acutely from an insufficient number of defection outcomes, we can examine the data in Table XI for their consistency with the models. Since the myopic and hyperopic models fail to agree in their attributions of bargaining advantages in only two cases, we exclude these cases from the table to simplify the exposition. In all the cases shown,

TABLE XI
Bargaining advantage and payoff allocation during defection from n-trial agreements to ally

Bargaining advantage attributed by:	Myopic and hyperopic models
Variables describing payoff to advantaged player:	
Average NS	0.54
Median NS	0.63
Proportion $NS > 0.5$	0.80 ($p = 0.0547$)
Proportion $NS \geqslant 0.5$	0.90 ($p = 0.0107$)
Number of defection trials (N)	(10)
Proportion of *excessive* payoffs in favor of advantaged player	2/4 (n.s.)
Proportion of *uneven* splits in favor of advantaged player	3/6 (n.s.)

it is the player excluded from the previous n-trial agreement who enjoys the bargaining advantage during the defection trial. While the data in the upper half of this table are consistent with our predictions, the average NS value is low and the bottom two entries, while of secondary importance, fail to discriminate between advantages and disadvantaged players.

Further inquiry into these data, however, reveals a very interesting pattern. Of the ten defection trials in which the models identify the same bargaining advantage, five outcomes initiate a new n-trial agreement while the other five do not. *All of the deviant cases for the bargaining advantage predictions occur on defection trials that initiate a new n-trial agreement.* We interpret these results to suggest that in breaking an n-trial agreement the previously excluded player exploits his bargaining advantage effectively, either to obtain a favorable division of payoff on the defection trial or to extract an n-trial agreement during which, as we suggest in our analysis of composite games, he is likely to obtain a highly favorable share of the coalition's payoff.

6. Conclusions

The models we outline in this paper differ from previous models of coalition behavior in certain respects that we regard as important. Rather than assuming that players have interval-scaled utility functions, perfect information, and unlimited computational facility, these models incorporate a model of individual decision processes based on ordinal preferences and heuristic rules-of-thumb guiding the bargaining strategies of each player. In calculating the possibilities in the situation and estimating the dynamics of the bargaining process, we assume actors recognize and take into account the calculations of the other players, adjusting their own actions accordingly until convergence is reached, yielding a constrained joint 'optimization' for members of the coalition. Thus, a model of bargaining processes is incorporated in the models. The models produce stochastic predictions as to what coalitions form and also predict the payoff allocations within the coalition. These predictions are quantitative, rather than the qualitative predictions typically made by behavioral models of coalition formation. One remaining feature is perhaps the most important of all to the models' potential relevance for empirical theory: the models focus on the interdependence of games rather than considering each episode in a continuing situation as a one-shot affair.

In conducting preliminary tests of the models, we analyze those pilot study outcomes dictated by an agreement applying only to a single trial. Each of the models receives strong and persistent support from these data. Although the pilot study generally provides too few cases in which the models make different predictions to place the models into effective competition, the data suggest that players act as if they consult both models in forming a coalition and dividing the payoff. We suggest a composite theory that incorporates the two models we present here, thereby assuming that players include both short and long-run consideration in developing their strategies.

If, indeed, players continue to form coalition agreements that dictate the outcome of only single trials, then the theories predict the life span of any coalition to be necessarily short. The probability that coalition (I, J) forms and persists n-trials may be computed from the expression $\prod_{t=1}^{n} P^*(I, J)_t$, the product of the appropriate $P^*(I, J)$ for each of the n-trials. Now as is the nature of the multinominal distributions across independent trials (except for interdependencies recognized by the models), for any $P^*(I, J)$ appreciably different from unity, the coalition is not likely to survive many trials without active intervention by the players. Such intervention appears in the pilot study (and in real-world political, social and economic life) in the guise of long-term agreements. Although our two models don't formally consider n-trial agreements, they nonetheless prove useful in explaining and predicting coalition behavior under the n-trial agreements in the pilot study. Using hypotheses derived from or suggested by these models, we enjoy success in predicting the timing, participants, and payoff allocations associated with the initiation, maintenance, and defection from n-trial agreements to ally.

Additional empirical and theoretical work is in order. Empirically, we are encouraged by our preliminary analyses of data from a replication of the pilot study in two different populations. Theoretically, the ease with which new features can be incorporated into the model suggest that this approach promises to be useful in developing theory increasingly able to recognize the complexity of coalition processes characteristic of continuing economic, social and political relationships.[14]

Carnegie-Mellon-University

NOTES

[1] An earlier version of this paper was read at the 1971 Annual Meeting of the American Political Science Association. The research reported in this paper is supported by National Science Foundation Grant 1–55409. We also thank Richard M. Cyert for supporting the initial research, and Peter H. Aranson, Kenneth E. Friend, Martin S. Geisel, Terry C. Gleason, Melvin J. Hinich, Gordon Lewis, Bernhardt Lieberman, Kenneth R. MacCrimmon, Timothy W. McGuire, Peter C. Ordeshook, Anatol Rapoport, Howard Rosenthal, and Gerald A. Thompson for their helpful comments at various stages in the project.

[2] For a more complete discussion of the tests using only single trial agreements from the pilot study, see Laing and Morrison (in press).

[3] These models have been incorporated into a computer program ('COAL 03') written in FORTRAN IV and available on request.

[4] This function is described as 'pseudo' because the 'values' of the various coalitions in its statement are not expressed in a commodity for which the players' utilities are assumed to be linear. Our models specifically assume that the utility of coalition outcomes to each of the players is *not* linear in the points which are the medium in which payoffs are awarded during trials in the sequential game used in the pilot study. Also note that the (pseudo-) characteristic function is not superadditive. Superadditivity requires that the value of the three-person coalition, $V(I_n) \geq \max_i V_i$. Luce and Raiffa require superadditivity in defining the characteristic function (1957, pp. 183f). For an example of the utilization of nonsuperadditive characteristic functions, see Rapoport (1970, pp. 83, 121).

The concept of 'the coalition of the whole' is difficult to operationalize in our study. Although three-person coalitions were not recognized (hence are operationally undefined), formation of a two-person coalition required a modicum of cooperation from the excluded player. As a suggestive parallel, even though a senator may lose on a given vote, nevertheless he obeys the rules of the Senate, if only to protect his chances of future success.

[5] The model assumes that the player will obey these assumptions *except under duress*. In an earlier paper (in press) we operationalized the extent to which player I is subject to duress on trial T as the number of rank positions he is expected to lose on trial T according to the theory if he is excluded from the coalition. Then, for each coalition (I, J) we predicted that the partner, say I, subject to the greater rank threat was more likely than J to violate one or more of Assumptions 2.1–2.3. The results strongly support this prediction, although the effects of rank threat and bargaining disadvantage are severely confounded.

Note that given Assumption 1.1, each trial in our sequential game is a three-person 'game of status' as conceptualized by Shubik (1971).

[6] The model assigns a value to X_{ij}^- such that this is the minimum payoff in the (I, J) coalition that enables I to attain the rank associated with X_{ij}^+ and also permits I to obey Assumptions 2.1 and 2.2.

[7] This 'probability matching' is in the spirit of the Ofshes' (1970) model of coalition choice.

[8] Conceptions of this type often involve interpersonal comparisons of utility. For a discussion of the relevance of this issue to our attributions of bargaining advantage, see Laing (1974).

[9] In order to check our results independently of the n-trial agreement coding decisions, we identified those 73 cases from the pilot study data in which the same coalition did not form on the immediately preceeding trial. This method guarantees exclusion of all but the first trial outcome in any n-trial agreement (although, of course, this technique also excludes outcomes which are *not* dictated by an n-trial agreement). This rule has the important disadvantage that it is *post hoc*, requiring us to admit cases after the fact, and therefore cannot be used in predicting coalition behavior in real time. However, it has the important advantages of being both unambiguously replicable and independent of our coding decisions which one might suspect as being biased in favor of the theory. The existence of such a bias would result in our finding stronger support for the theory than were the bias not present. In fact, no such bias can be detected: the data from the 73 cases selected independently of our n-trial agreement coding decisions offer support for the theory which is essentially identical (or, if different, stronger) to that provided by the data selected as being 'clean' of n-trial agreements. Since it makes essentially no difference which data we use, we shall report findings based on the 76 cases identified *a priori* as uncontaminated by n-trial agreements.

[10] A number of theories in the social sciences assume that profitable interactions tend to recur. It is interesting that if we predict for the 76 cases that the coalition that formed on the previous trial forms again, we are correct in *less than $\frac{1}{3}$ of the cases*. That is, the persistence hypothesis enjoys less success than the null hypothesis that each coalition is equally likely.

[11] Asymptotically, 2 times the natural logarithm of the likelihood ratio has a χ^2 (1) distribution, but the asymptote is frequently approached rather slowly. We regard 76 cases as too small a number for using the test with confidence; however if we apply the test willy nilly, both the myopic and hyperopic models perform significantly better than the null model ($p < 0.05$).

[12] Given the location of each case on the (antecedent) predictor variable, the decision rule if 'agree' then predict 'current trial only' (or, equivalently, if 'n-trials' then predict 'disagree') provides a 43% reduction in errors of prediction over those committed

under the null hypothesis. For a development of the prediction language, measures of association custom designed for specific hypothesis stated in the prediction language, and asymptotic sampling theory for the general $R \times C$ bivariate case, see Hildebrand, Laing, and Rosenthal (in press).

[13] See note 5.

[14] For an extended discussion of the theoretical and methodological implications of the general approach used in developing the models discussed in this paper, see Laing et al. (1973).

BIBLIOGRAPHY

Emerson, Richard M.: 1962, 'Power-Dependence Relations', *American Sociological Review* **17** 31–41.

Emerson, Richard M.: 1964, 'Power-Dependence Relations: Two Experiments', *Sociometry* **27** 282–298.

Festinger, Leon: 1954, 'A Theory of Social Comparison Processes', *Human Relations* **7** 117–140.

Hildebrand, David K., Laing, James D. and Rosenthal, Howard: 1973, 'Prediction Logic: A Method for Empirical Evaluation of Formal Theory', *Journal of Mathematical Sociology* **3** (in press).

Hoffman, Paul J., Festinger, Leon and Lawrence, Douglas H.: 1954, 'Tendencies Toward Group Comparability in Competitive Bargaining', *Human Relations* **7** 141–159.

Kaplan, Morton: 1957, *System and Process in International Politics*, Wiley, New York.

Laing, James D.: 1974, 'Power, Dependence and Interpersonal Comparisons of Utility in *n*-Person Supergames', *Contemporary Management: Issues and Viewpoints*, Chap. 14 (ed. by Joseph W. McGuire), Prentice Hall, Englewood Cliffs, N.Y.

Laing, James D. and Morrison, Richard J.: 'Coalitions and Payoffs in Three-Person Sequential Games: Initial Tests of Two Formal Models', *Journal of Mathematical Sociology* **3** (in press).

Laing, James D. and Morrison, Richard J.: 1971, 'Coalition Formation in Certain Sequential Three-Person Games. III: Coalition Stability and *n*-Trial Agreements', Carnegie-Mellon University, Graduate School of Industrial Administration Working Paper WP 87-70-1.

Laing, James D., Lebanon, Alexander, Morrison, Richard J. and Rosenthal, Howard: 1973, Computer Models of Political Coalitions', *Political Science Annual* IV, (ed. by Cornelius Cotter), Bobbs-Merill, in press.

Luce, R. Duncan and Raiffa, Howard: 1957, *Games and Decisions*, Wiley, New York.

Ofshe, Lynne and Ofshe, Richard: 1970, *Utility and Choice in Social Interaction*, Prentice-Hall, Englewood Cliffs, N.J.

Rapoport, Anatol: 1970, *n-Person Game Theory*, University of Michigan Press, Ann Arbor.

Shubik, Martin: 1971, 'Games of Status', *Behavioral Science* **16** 117–129.

Vinacke, W. Edgar: 1959, 'The Effect of Cumulative Score on Coalition Formation in Triads with Various Patterns of Internal Power' (Abstract), *American Psychologist* **14**, 391.

Vinacke, W. Edgar: 1962, 'Power, Strategy and the Formation of Coalitions in Triads under Four Incentive Conditions', Office of Naval Research, NONR 3748 (02) Technical Report No. 1.

M. FREIMER AND P. L. YU

THE APPLICATION OF COMPROMISE SOLUTIONS TO REPORTING GAMES

ABSTRACT. This is a continuation of Yu's previous work on the compromise scheme for group decision problems with a new application to reporting games [12]. We first show that the distance parameter p for group regret has the interesting property that as p is increased, individual regrets are more emphasized and the sum of individual utilities resulting from the compromise scheme is decreased. In particular, when $p=1$, the sum of individual utilities is most emphasized and individual regret is most neglected. We then set up reporting games in the compromise scheme and show the existence of Nash equilibrium point. Finally we introduce new concepts of 'semioptimality' and 'optimality' to study two special classes of reporting games. We explicitly state and prove the semioptimal and optimal strategies for each player. We also show that 'simple majority rule' is a special case of compromise solutions with parameter $p=1$, which once again is the one which most neglects the individual regrets in the entire family of compromise solutions.

1. INTRODUCTION

Suppose that there are n persons who engage in a joint business (say a partnership or a corporation). It is not unusual that they may have different opinions on decision making, even though they have a common goal. When a disagreement occurs, how do they compromise in such a way as to enforce a genuine cooperation among the group so that their overall objectives may be maximally achieved? This kind of group decision problem has been studied for a long time [1]-[11], although there are still no universally accepted solution concepts. Recently, Yu [12] proposed a class of compromise solutions for the problem and investigated its merits and weaknesses. He showed that, although it is by no means perfect, the solution concept does enjoy some nice properties such as feasibility, individual rationality, least group regret, no dictatorship, Pareto optimality, uniqueness, symmetry (equity principle) and independence of irrelevant alternatives. As pointed out by Professor A. Rapoport [13], there are a few open questions in the compromise solutions which are not explored in [12]. In this note, we try to answer some of these open questions. In order to be precise, let us summarize some relevant results of [12] as follows.

Suppose there is a group of n-persons, each of them having a utility

function $u_j(x), j=1,\ldots,n$, defined over X; X is the set of feasible alternatives for a decision to be made by the group. We shall assume that X is compact and each $u_j(x)$ is continuous.
Let $u(x)=(u_1(x),\ldots u_n(x))$,

$$X_0 = \{x \in X \mid u(x) \geq 0\}[1] \quad \text{and} \quad u_j^* = \max\{u_j(x) \mid x \in X_0\}$$

which is assumed to exist for all $j=1,\ldots,n$. Then

$$U_0 = \{u(x) \mid x \in X_0\}$$

is the set of all feasible outcomes in the utility space and $u^* = (u_1^*, \ldots, u_n^*)$ is the *utopia point* of our problem, because if u^* is obtainable then the group will unanimously select u^* for their decision. Now for $p \geq 1$, given a feasible decision $x \in X_0$, we associate it with a *group regret* by $D_p(u(x)) = = [\sum_{i=1}^{n}(u^* - u_i(x))^p]^{1/p}$. Note that the group regret of a decision x is represented by the distance between the utopia point and $u(x)$ in the utility space.

DEFINITION 1.1. The decision $x_0 \in X_0$ and its outcome $u(x_0)$ are called the *compromise solution* with parameter $p \geq 1$ if x_0 minimizes $D_p(u(x))$ over X_0. For notational convenience, the compromise solution with parameter p in the utility space will be denoted by $u(p) = (u_1(p), \ldots, u_n(p))$, (instead of u^p in [12]). The properties of $u(p)$ as mentioned above have been explored in [12]. The bounds and monotonicity of $u(p)$ in terms of p, when $n=2$, is also studied in [12].

In this paper we shall investigate some important problems which have not been studied before. The first question is what is the group impact of p on the compromise solution (in contrast to the impact to the individual participant)? We shall investigate this problem in Section 2. Our main result is that as p is increased, individual regrets are more emphasized and the sum of individual utilities resulting from the compromise scheme is decreased.

The next main question we want to tackle is that, if each $u_j(x)$ is not reported truthfully, what is the impact on the compromise scheme. We shall study this question is Section 3. In particular in Section 3.1., we set up a general model for the reporting games and show the existence of a Nash equilibrium point. We also introduce a concept of *semioptimality* which enjoys the property of individual stability in selecting strategies, and a concept of optimality which enjoys the property of both individual and group stability. Following the concept of optimality, whenever unique optimal strategies exist the value of the games to each player is determined. In Section 3.2. we study n-person two choices cases. We derive the optimal strategies and the values of games for each player and show how majority rule in simple voting games is a special case of our reporting game and also how individual regret is most neglected in our entire class of reporting games in the compromise scheme. Finally we study the semioptimal and optimal strategies in the reporting games with two persons and three choices in Section 3.3. We derive some easily applied pure semioptimal and optimal strategies for each player.

2. THE IMPACT OF PARAMETER p ON COMPROMISING SCHEME

In this section we shall study the effect of p on the sum of all individual utility resulting from the compromise scheme. This information may enable us to pick a suitable p for the compromise scheme. We shall assume in this section that the reported utilities are all truthful and limit ourselves to $n=2$, because in this case our result could be easily derived from the previous result of [12]. For more general cases which involve more technical development we refer to [14]. Now suppose U_0 is convex.[2] Then all compromise solutions, $u(p)$, $1 \leq p < \infty$, are Pareto optimal [12]. Without loss of much generality, the set of all Pareto solutions can be represented by

$$W = \{(u_1, u_2) \mid u_2 = g(u_1), u_1^0 \leq u_1 \leq u_1^*\}$$

where $u_1^0 = g^{-1}(u_2^*)$ and $g(u_1)$ is differentiable, concave and a strictly decreasing function (see Figure 1). The compromise solutions[3] $u(1)$ and $u(\infty)$ are depicted as in Figure 1. From [12] we know that all other compromise solutions $u(p)$, $1 < p < \infty$, lie on W monotonically from $u(1)$ to $u(\infty)$.

We are interested in how $u_1(p) + u_2(p)$ behaves as a function of p. Since for all Pareto optimal points $u_2 = g(u_1)$, we first consider the function $u_1 + g(u_1)$ for $u_1^0 \leq u_1 \leq u_1^*$. Since the point $u(1)$ minimizes $(u_1^* - u_1) + (u_2^* - u_2)$ over U_0, it maximizes $u_1 + u_2$ over U_0. Hence, $u_1(1)$ maximizes $u_1 + g(u_1)$. Furthermore $g(u_1)$ is concave,[4] so that $u_1 + g(u_1)$ is an in-

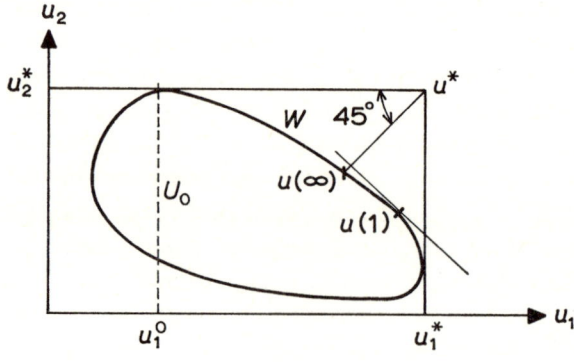

Fig. 1.

creasing function if $u_1^0 \leq u_1 \leq u_1(1)$ and is a decreasing function if $u_1(1) \leq u_1 \leq u_1^*$. Since $u_1(p)$ varies monotonically from $u_1(1)$ to $u_1(\infty)$ as p varies from 1 to ∞, we have

THEOREM 2.1. *Suppose $g(u_1)$ is differentiable and concave over $[u_1^0, u_1^*]$. Then $u_1(p) + u_2(p)$ is a decreasing function of p.*

REMARK 2.2. As in [12], the condition that $g(u_1)$ is differentiable could be relaxed to being piecewise differentiable. The theorem states that the larger is p, the smaller is the sum of all individual utilities resulting from the compromise scheme. Observe that, when $p=1$ the compromise solution minimizes the sum of the individual regrets; equivalently it maximizes the sum of the individual utilities. When $p=\infty$, the compromise solution minimizes the maximum of the individual regrets (because $D_\infty(u) = = \max_i \{u_i^* - u_i\}$). It is not unexpected that $\sum_i u_i(\infty) \leq \sum_i u_i(1)$. From these, we see intuitively that the larger is p, the greater is the emphasis on the individual regret and the smaller is the sum of individual utilities resulting from the compromise scheme.

The above intriguing property of p could also be observed from the family of isovalued curves of p. In Figure 2, I_1, I_2 and I_∞ represent the isovalued curve of $D_p(u)$ for $p=1$, 2 and ∞ respectively. We see that along I_1, the rate of substitution of the individual regrets is a constant, independent of where is the point of status quo. Thus each individual's regret is entirely submerged in the group. We also observe how inflexible is

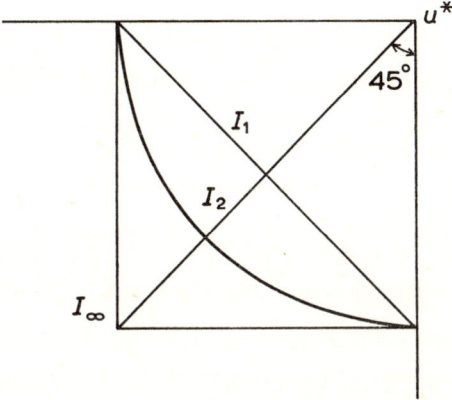

Fig. 2.

I_∞; along I_∞ substitution of the individual regrets is impossible. The group always registers the maximum of individual regrets as the group regret. None of the individual regret is neglected. As we increase p from 1 to ∞, we see that the property of the rate of substitution of individual regret moves from that of $p=1$ to that of $p=\infty$. Although each $D_p(u), p \geq 1$, indicates a distance of u to the utopia point u^*, we see how p plays the important role of determining how we balance individual regret and that of the group as a whole. (For a further discussion, see Remark 3.2.3.)

3. Reporting Games within Compromising Schemes

In this section we shall remove the assumption that the utility $u(x)$ is truthfully reported. Instead, we shall assume that for each player j, $u_j(x)$ is partially known. Also for each $x \in X$, there is a *credit interval*[5] $I(j, x)$ for each player j with an interpretation that the jth player's reporting utility of x must lie in the set $I(j, x)$; otherwise he will lose his creditability which would be a disaster to him (no one will trust him anymore, say). We shall assume that $\{I(j, x) \mid j \in N, x \in X\}$ where $N = \{1, ..., n\}$ are known to each participant. Suppose all participants agree to use a compromise scheme to resolve their different opinion in decision making. How do they report $u_j(x)$ so as to maximize their own utility?

3.1. *The Format of the Reporting Games*

In order to facilitate our presentation, we shall assume that X is a finite set. In particular we set $X = \{1, 2, \ldots, q\}$. Thus q is the number of feasible choices under consideration.

First let
$$I(j) = I(j, 1) \times \cdots \times I(j, q) \quad j = 1, \ldots, n$$
and
$$J(x) = I(1, x) \times \cdots \times I(n, x) \quad x = 1, \ldots, q.$$

We see that $I(j)$ is a q-dimensional rectangle which is the set of all feasible pure strategies enjoyed by the jth participant. On the other hand, $J(x)$ is an n-dimensional rectangle which represents the set of all feasible reporting utilities for the choice x.

Let $r_j(x)$ be the jth player's reporting utility for the choice x, and let $r(x) = (r_1(x), \ldots, r_n(x))$. From our assumption, it is seen that $r(x) \in J(x)$. Denote the set of all reporting points $r(x)$ by R. That is, $R = \{r(x) \mid x \in X\}$. We also denote the convex hull of R by $C[R]$. That is,

$$C[R] = \left\{ \sum_{x \in X} \lambda(x) r(x) \,\Big|\, \lambda(x) \geq 0 \atop \sum_{x \in X} \lambda(x) = 1 \right\}$$

As an example, see Figure 3, we have a problem of two players and three choices. The rectangles $J(1)$, $J(2)$, and $J(3)$ represent the set of possible reporting points for choice 1, 2, and 3 respectively. Suppose the players report their utilities as $r(1)$, $r(2)$ and $r(3)$. Then R is the set which contains those three points $r(x)$; while $C[R]$ is the triangle with vertices at $r(1)$, $r(2)$ and $r(3)$.

Now given the set of reporting points R, instead of U_0 in the truthful case, we shall consider $C[R]$ as the *reported utility space* for the compromise scheme. And instead of u^*, we shall consider the *reported utopia point* $r^*(R) = (r_1^*, \ldots, r_n^*)$ where $r_j^* = \max\{r_j(x) \mid x \in X\}$.

DEFINITION 3.1.1. $r^0(R; p) \in C[R]$ is a compromise solution in the reported utility space with parameter $p \geq 1$, if it minimizes

$$D_p(R, r) = \left[\sum_{j=1}^{n} (r_j^* - r_j)^p \right]^{1/P} = \|r^* - r\|_p \text{ for all } r \in C[R].$$

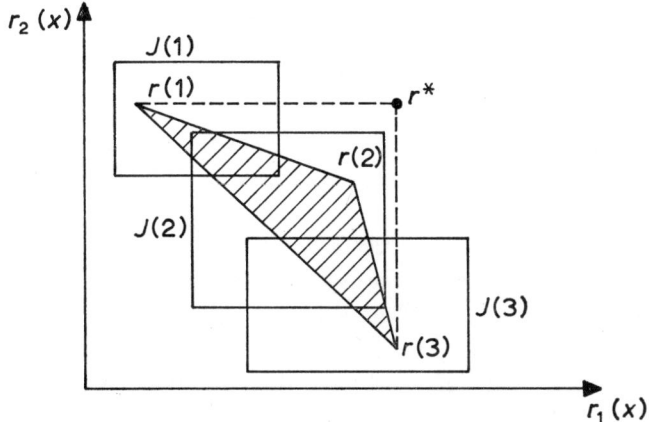

Fig. 3.

REMARK 3.1.2. The compromise solution in the reported utility space has the same structure as that in the truthful utility space.

Since $r^0(R; p) \in C[R]$, we can write

(1) $\quad r^0(R; p) = \sum_{x \in X} \lambda_x^0(R; p) r(x)$ with

$\lambda_x^0(R; p) \geq 0$ and $\sum_{x \in X} \lambda_x^0(R; p) = 1$.

Suppose the representation of $r^0(R; p)$ in terms of $\lambda_x^0(R; P)$ is unique in (1). We can then define the payoff to the jth player with reporting strategies R and compromise parameter p by

(2) $\quad P_j(R; p) = \sum_{x \in X} \lambda_x^0(R; p) u_j(x), \quad j = 1, 2, \ldots, n$

where $u_j(x)$ is the true utility of x to the jth player.

REMARK 3.1.3. Suppose that the representation of $r^0(R; p)$ in terms of $\lambda_x^0(R; p)$ in (1) is not unique.[6] Then the payoffs $P_j(R; p)$ in (2) are not uniquely defined. Instead of making certain conventions to enforce uniqueness, we shall assume from now on that for each R and p the representation of $r^0(R; p)$ in terms of $\lambda_x^0(R; p)$ is unique. In our next sections, it can easily be checked that this assumption is satisfied.

REMARK 3.1.4. In the definition of our payoffs, we implicitly assume that each choice is divisible. Observe that, because $\lambda_x^0(R;p) \geq 0$ and $\sum_{x \in X} \lambda_x^0(R;p) = 1$, each $\lambda_x^0(R;p)$ could be interpreted as a probability mass as well as the relative importance each x should be given. For instance, suppose we have a police force of 100 persons which is going to be divided into two policing districts. Suppose $\lambda_1^0(R;p) = 0.4$ and $\lambda_2^0(R;p) = 0.6$. We may put 40 persons and 60 persons in district 1 and 2 respectively. Note that in this example the reporting utility in x would be the utilities if all 100 persons are put in the district x.

LEMMA 3.1.5. *Suppose that the assumption of the unique representation in Remark 3.1.3 is satisfied. Then $P_j(R;p)$ is a continuous function of R for each $j = 1, \ldots, n$ and $p > 1$.*

Proof. See Appendix (for the sake of a smooth presentation.) Following the result of Glicksberg [16], we have

THEOREM 3.1.6. *Suppose that each credit interval $I(i, x)$ is compact (thus the set of all pure strategies $I(i)$ is compact) and that the unique representation assumption in Remark 3.1.3 is satisfied. Then our reporting game has at least one Nash equilibrium point (in the sense of mixed strategies).*

REMARK 3.1.7. Although a Nash equilibrium point exists for our reporting game, to actually find it is a nontrivial job. Because there are quite a few defects in the solution concept of Nash equilibrium point [5, 7, 8] we do not think that we shall devote our effort to find the Nash equilibrium points in our general reporting games. Rather, we shall limit ourselves to some special classes of reporting games which have pure 'optimal' strategies for each player. Our goal is to specify those optimal strategies so that they are readily applicable. In order to facilitate our latter discussion, we need two concepts. Recall that $R = \{r(x) \mid x \in X\}$ $X = \{1, \ldots, q\}$ and each $r(x) = (r_1(x), \ldots, r_n(x))$. We may consider R as a set of $n \times q$ decision variables. Given $r_j(x)$, by $R - r_j(x)$ we mean the set of the remaining decision variables of R after $r_j(x)$ is removed. When we are considering a particular decision variable $r_j(x)$, we shall rewrite the jth player's payoff as $P_j(r_j(x), R'; p)$ where $R' = R - r_j(x)$. Observe that by our notation

(2)′ $\quad P_j(r_j(x), R'; p) = P_j(R; p).$

DEFINITION 3.1.8. *Given $r_j^0(x) \in I(j, x)$, we say that $r_j^0(x)$ is a semioptimal* *control* for the jth player if

(2)″ $\quad P_j(r_j^0(x), R'; p) \geq P_j(r_j(x), R'; p)$

for all $r_j(x) \in I(j, x)$ and for all fixed R'. Thus $r_j^0(x)$ is semioptimal for the $j$the player if, no matter what are the choices for $R - r_j(x)$, $r_j^0(x)$ always maximizes the resulting payoff of the jth player.

The definition of semioptimal control can be easily extended to *semioptimal strategies*.

DEFINITION 3.1.9. A strategy μ_j for the jth player, is semioptimal if, no matter what are the choices of the other players' strategies, the strategy μ_j always maximizes the resulting payoff of the jth player.

DEFINITION 3.1.10. Let $\mu = (\mu_1, ..., \mu_n)$ be a set of strategies for the n players. We say μ is semioptimal if each μ_j is semioptimal; and μ is *optimal* if it is semioptimal and also Pareto optimal.

REMARK 3.1.11. Semioptimality has the property of individual stability, because the strategy always is the best for one player against any other player's strategies. On the other hand Pareto optimality has a group stability; any set of strategies which is not Pareto optimal could provoke a group negotiation to improve their common welfare. We see that our definition of optimal strategies does have individual and group stabilities in the sense described above.

REMARK 3.1.12. In two person zero-sum-games, a pair of semioptimal strategies (if it exists) forms an equilibrium point. The converse is not generally true. As an example, in the following matrix game, (B, A) is an equilibrium point.

$$\begin{array}{c} & A & B \\ A & \begin{bmatrix} 2 & 0 \\ 3 & 5 \end{bmatrix} \\ B & \end{array}$$

But A is not a semioptimal strategy for the second player (if Player 1, by mistake, uses A, then Player 2 should use B to capitalize the mistake). This shows that semioptimality is a more restrictive concept than that of equilibrium point.

REMARK 3.1.13. As the following bimatrix game (the 'prisoner's dilemma') shows, semioptimal strategies are not necessarily Pareto optimal, and optimal strategies need not exist.

$$\begin{array}{c c} & \begin{array}{c c} A & B \end{array} \\ \begin{array}{c} A \\ B \end{array} & \left[\begin{array}{c c} (4, 4) & (0, 5) \\ (5, 0) & (1, 1) \end{array} \right] \end{array}$$

Here (B, B) is the only semioptimal strategy, but it is not Pareto optimal. On the other hand, (A, A) is Pareto optimal but is not semioptimal.

REMARK 3.1.14. Any semioptimal strategy is clearly a Nash equilibrium point. The converse is not generally true. For example, in the following bimatrix game

$$\begin{array}{c c} & \begin{array}{c c} A & B \end{array} \\ \begin{array}{c} A \\ B \end{array} & \left[\begin{array}{c c} (5, 1) & (0, 0) \\ (0, 0) & (1, 5) \end{array} \right] \end{array}$$

both (A, A) and (B, B) are Nash equilibrium points but neither A nor B is a semioptimal strategy for either player. In fact, there are no semioptimal strategies in this game. In the sequel, we need the following definition.

DEFINITION 3.1.15[7]. Suppose the set of optimal strategies is unique. The resulting payoff to the jth player will be called the *value of the game* to the jth player.

In the next two sections, we shall describe some semi-optimal and optimal strategies for some classes of our reporting games. Observe that the value of our game is defined directly from the unique set of optimal strategies (when they exist), rather than from the characteristic function of the game. This makes it possible that our value of a game (when it exists) is different from the Shapley value. In order to see this point, let us consider a three person game. Each player has two choices A and B. The payoff to the players is given by

Player 3 uses A   Player 3 uses B

Player 2   Player 2

$$\begin{array}{c c} & \begin{array}{c c} A & B \end{array} \\ \text{Player 1} \quad \begin{array}{c} A \\ B \end{array} & \left[\begin{array}{c c} (5, 5, 3) & (3, 3, 3) \\ (2, 2, 3) & (1, 1, 3) \end{array} \right] \end{array} \qquad \begin{array}{c c} & \begin{array}{c c} A & B \end{array} \\ \text{Player 1} \quad \begin{array}{c} A \\ B \end{array} & \left[\begin{array}{c c} (3, 3, 2) & (2, 2, 2) \\ (1, 1, 2) & (0, 0, 2) \end{array} \right] \end{array}$$

We see that (A, A, A) is the optimal strategy with value (payoff) $(5, 5, 3)$ to Player 1, 2, 3 respectively. Now let us use 'maximin' criterion to derive the characteristic function of the game as follows:

$$V(\phi) = 0, \quad V(1) = 2, \quad V(2) = 1, \quad V(3) = 3, \quad V(1, 2) = 6$$

(the sum of the first two entries of $(3, 3, 2)$ which is the payoff of (A, A, B); we assume that the individual payoffs are summable.)

$$V(1, 3) = 6, \quad V(2, 3) = 5, \quad V(1, 2, 3) = 13.$$

The Shapley value of this game is given by:

$$\psi(1) = 1/3(2) + 1/6(5 + 3) + 1/3(8) = 14/3$$
$$\psi(2) = 1/3(1) + 1/6(4 + 2) + 1/3(7) = 11/3$$
$$\psi(3) = 1/3(3) + 1/6(4 + 4) + 1/3(7) = 14/3$$

We see that our value of this game is different from Shapley's value. Observe that our value is a member of the core of the game. The reason for the difference between our value and Shapley's value together with their applicability is currently under investigation.

3.2. N-Person Reporting Games with Two Choices

We shall assume that each credit interval $I(j, x), j=1,...,n$ $x=1, 2$, is compact and let $\bar{I}(j, x)$ and $\underline{I}(j, x)$ respectively be the least upper bound and the greatest lower bound of $I(j, x)$. Given x, we shall denote the other choice by x'. Our main result is

THEOREM 3.2.1. *Optimal reporting strategies exist and are given by*

$$r_j(x) = \begin{cases} \bar{I}(j, x) & \text{if } u_j(x) > u_j(x') \quad \text{(i.e., } x \text{ is preferred to } x') \\ \underline{I}(j, x) & \text{if } u_j(x) < u_j(x') \\ \text{arbitrary} & \text{if } u_j(x) = u_j(x') \end{cases}$$

$x = 1, 2; \quad j = 1,...m$ n.

We need the following lemma for our theorem.

LEMMA 3.2.2. *Let* $R = \{r(1), r(2)\}$ *be the set of two reporting points with* $r(1) = (r_1(1),...,r_n(1))$ *and* $r(2) = (r_1(2),...,r_n(2))$. *Then the compromise solution in the reported utility space is given by* $r^0(R, p) = r(1) +$

$+\theta^0(R,p)(r(2)-r(1))$ where

$$\theta^0(R,p) = \frac{[\sum_{j\in B}(r_j(2)-r_j(1))^p]^{1/p-1}}{[\sum_{j\in A}(r_j(1)-r_j(2))^p]^{1/p-1}+[\sum_{j\in B}(r_j(2)-r_j(1))^p]^{1/p-1}}$$

and

$$A = \{j \mid r_j(1) \geq r_j(2)\}$$
$$B = \{j \mid r_j(2) > r_j(1)\}.$$

Proof. By definition we have

(3) $\quad r_j^* = \begin{array}{l} r_j(1) \text{ if } j\in A \\ r_j(2) \text{ if } j\in B. \end{array}$

Note, we can write

$$C[R] = \{r(1) + \theta(r(2)-r(1)) \mid 0 \leq \theta \leq 1\}.$$

Thus $r^0(R,p)$ minimizes

$$D^p(R,r) = \left[\sum_{j=1}^n (r_j^* - r_j)^p\right]^{1/p}$$

where

(4) $\quad r = r(1) + \theta(r(2) - r(1)), \quad 0 \leq \theta \leq 1.$

Note that $r^0(R,p)$ also minimizes

(5) $\quad \bar{D}_p(R,r) = \sum_{j=1}^n (r_j^* - r_j)^p$

for those r satisfying (4).

Putting (4) in (5), we get from (3),

(6) $\quad \bar{D}_p(R,r) = \sum_{j\in A}[r_j(1) - r_j(1) - \theta(r_j(2)-r_j(1))]^p +$
$\quad\quad + \sum_{j\in B}[r_j(2) - r_j(1) - \theta(r_j(2)-r_j(1))]^p =$
$\quad\quad = \theta^p \sum_{j\in A}(r_j(1)-r_j(2))^p + (1-\theta)^p \sum_{j\in B}(r_j(2)$
$\quad\quad\quad\quad\quad - r_j(1))^p.$

By taking the derivative of (6), one finds that $\theta^0(R,p)$ as defined in the lemma minimizes $\bar{D}_p(R,r)$ over $r \in C[R]$.

Q.E.D.

Proof of Theorem 3.2.1. From (2) of Section 3.1 and Lemma 3.2.2, we know that the payoff to the jth player with respect to R and p is given by

$$P_j(R;p) = (1 - \theta^0(R,p))\,u_j(1) + \theta^0(R,p)\,u_j(2).$$

Suppose $u_j(1) > u_j(2)$. Then the jth player wants to decrease $\theta^0(R;p)$ as much as possible in order to increase his payoff $P_j(R;p)$. From the expression of $\theta^0(R,p)$ in Lemma 3.2.2, we see that this can be achieved by setting $r_j(1) = I(j, 1)$ and $r_j(2) = \bar{I}(j, 2)$. Similarly, if $u_j(1) < u_j(2)$, the jth player can set $r_j(1) = \bar{I}(j, 1)$ and $r_j(2) = I(j, 2)$ to maximize his payoff. This proves the semioptimality. The Pareto optimality is clear. The remaining case that $u_j(1) = u_j(2)$ is trivial.

Q.E.D.

Next let us investigate the relationship between the *rule of simple majority vote* and compromise solutions. We will show that the rule of simple majority vote essentially is a compromise solution with parameter $p=1$. In order to see this, let each credit interval contain only two points. More precisely, let $I(j, x) = \{1, 0\}$ and thus $r_j(x) = 1$ (or 0) if the jth player reports favoring (or unfavoring) the alternative x. Let us assume that those players who are indifferent to the two choices should abstain. Thus, we shall interpret n in our compromising scheme as the number of players who actually vote and show their preference between the choice 1 or 2. Thus $r_j(1) \neq r_j(2)$ and both of them can be only 0 or 1. For latter reference, we call group decision problems of this type *simple voting games*.

According to Theorem 3.2.1, the optimal reporting strategies for the jth player, $j = 1, \ldots, n$, is

(7) $\quad$ if $\;u_j(1) > u_j(2)\;$ then $\;r_j(1) = 1\;$ and $\;r_j(2) = 0;$
$\quad\quad\;\;$ if $\;u_j(2) > u_j(1)\;$ then $\;r_j(1) = 0\;$ and $\;r_j(2) = 1.$

Under the optimal play, the definition of A and B becomes

$$A = \{j \mid u_j(1) > u_j(2)\} \quad \text{(those who prefer 1 to 2)}$$
$$B = \{j \mid u_j(2) > u_j(1)\} \quad \text{(those who prefer 2 to 1)}$$

and

(8) $$\theta^0(p) = \frac{|B|^{1/p-1}}{|A|^{1/p-1} + |B|^{1/p-1}}$$

where $|A|$ and $|B|$ are the numbers of the elements in A and B respectively.

The value of the game (the payoffs under the optimal play) are

(9) $\quad u_j(p) = (1 - \theta^\circ(p))\, u_j(1) + \theta^\circ(p)\, u_j(2) \qquad j = 1, \ldots, n$

where $\theta^\circ(p)$ is given in (8).

Observe that if either A or B is empty, then $\theta^\circ(p)$ is either 1 or 0, and the optimal group decision is unanimous (either 2 or 1). Now suppose that neither A nor B is empty.

Then

(10) $\quad \theta^\circ(p) = 1/2 \quad \text{for all} \quad p \geq 1 \quad \text{if} \quad |A| = |B|$

(11) $\quad \lim_{p \to 1} \theta^\circ(p) = \begin{cases} 1 & \text{if } |B| > |A| \\ 0 & \text{if } |A| > |B| \end{cases}$

(12) $\quad \lim_{p \to \infty} \theta^\circ(p) = 1/2.$

REMARK 3.2.3. For $p = 1$, from (11), we see that if $|B| > |A|$ then the group will take choice 2, and that if $|A| > |B|$ then the group will take choice 1. This is exactly a simple majority rule. On the other hand, when $p = \infty$, we see that whenever unanimous decision is impossible (i.e., neither A or B is empty) the group's compromise solution should be half of each choice no matter how many people are in favor of A or B. Recall from Remark 2.2, that when $p = \infty$ the individual regrets are most emphasized and when $p = 1$ the sum of the individual utilities are most emphasized. In other words p could be regarded as an indicator of how much emphasis is put on each individual and how much emphasis is put on the group. In our simple voting games, it turns out that the rule of simple majority vote coincides with our compromise solution with $p = 1$. Thus, the rule of simple majority vote is one which most emphasizes the group and thus most neglects the individual in our entire family of compromising schemes.

We summarize the highlights of the above discussion into

THEOREM 3.2.4. In the simple voting games, the optimal strategies and the values of the games are given in (7) and (9) respectively. When $p = 1$, the compromise scheme coincides with the simple majority rule; and when $p = \infty$, if unanimous decision is impossible the compromise scheme always yields half of each of the two choices no matter how many more people are in favor of one choice rather than the other.

3.3. Two Person Reporting Games with Three Choices

In this section we shall derive a set of pure optimal strategies for both players when our reporting games satisfy some sufficiency conditions. In order to achieve this goal we shall assume in this section that, for each $j=1$ or 2, $I(j, x) > I(j, x')$ implies that $u_j(x) > u_j(x')$. This assumption states that if the credit interval of x has a higher upper limit than that of x', then the true utility of x is higher than that of x'. We shall also assume that for each j

$$I(j, x) \neq I(j, x') \quad \text{if} \quad x \neq x'.$$

Thus implicitly we have assumed that for each player there is only one $x(j)$ which yields the highest utility. Once it is revealed that $u_j(x) > u_j(x')$, we shall assume that $r_j(x) > r_j(x')$. (Otherwise, it is an obvious cheating.) Without loss of generality (renumber the index if necessary), let

$$I(1, 1) < I(1, 2) < I(1, 3) \quad \text{and}$$
$$I(2, x_1) < I(2, x_2) < I(2, x_3).$$

Suppose that $x_3 = 3$. From our assumption the choice 3 will be the unanimous decision for the group. No problem could possibly arise. Thus from now on we shall assume that $x_3 \neq 3$. In particular we set $x_3 = 1$. According to the notation of Section 3.1, we have three rectangles

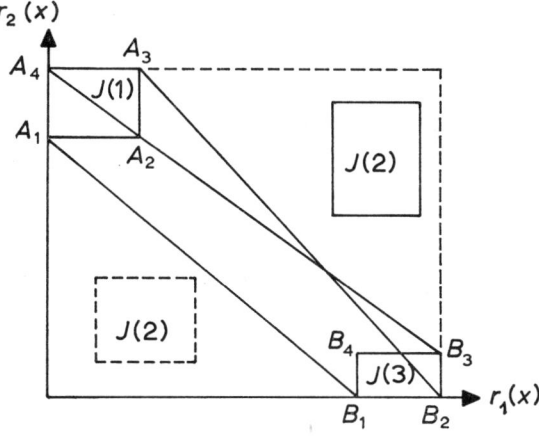

Fig. 4.

$J(x)$, $x=1, 2, 3$, which represent the sets of all feasible reporting points of x. In Figure 4 we represent $J(1)$ and $J(3)$ by the rectangles $[A_1, A_2, A_3, A_4]$ and $[B_1, B_2, B_3, B_4]$ respectively. We have the following result.

THEOREM 3.3.1. (See Figure 4). Suppose $J(2)$ lies under the line $[A_1, B_1]$. Then for all $p \geq 1$, the following strategies are optimal:

$$r_1(x) = \begin{cases} I(1, 1) \text{ for } x = 1 \\ \text{arbitrary}, \ x = 2 \\ I(1, 3), \quad x = 3 \end{cases}$$

$$r_2(x) = \begin{cases} I(2, 1) \text{ for } x = 1 \\ \text{arbitrary}, \ x = 2 \\ I(2, 3) \quad x = 3. \end{cases}$$

Proof. Under our assumption, no matter what is the reported control $r(2) \in J(2)$, $r(2)$ cannot affect the compromise solution (because a compromise solution must be a Pareto optimal in the reported utility space.) Our assertion follows immediately from Theorem 3.2.1.

THEOREM 3.3.2. (See Figure 4).

(i) Suppose $J(2)$ lies on top of the lines $[A_4, B_3]$ and $[A_3, B_2]$. Then for all $p \geq 1$, the following controls are semioptimal:

$$r_1(x) = \begin{cases} I(1, 1) \text{ for } x = 1 \\ I(1, 3) \quad x = 3 \end{cases}$$

$$r_2(x) = \begin{cases} I(2, 1) \text{ for } x = 1 \\ I(2, 3) \quad x = 3. \end{cases}$$

(ii) Suppose also that

$$I(1, 2) + I(2, 2) > \max\{I(1, 3) + I(2, 3), I(2, 1) + I(1, 1)\}.$$

Then for $p \geq 2$, the following pure strategies are semioptimal.

$$\mu_1 = (I(1, 1), I(1, 2), I(1, 3)) = (r_1(1), r_1(2), r_1(3))$$
$$\mu_2 = (I(2, 1), I(2, 2), I(2, 3)) = (r_2(1), r_2(2), r_2(3)).$$

(iii) When the assumptions in (i) and (ii) are satisfied and $J(2)$ lies on top of the line $[A_3, B_3]$, the strategies specified in (i) and (ii) are also optimal.

Proof. For (i) and (ii), we shall prove that $r_1(x)$, $x = 1, 3$ is semiop-

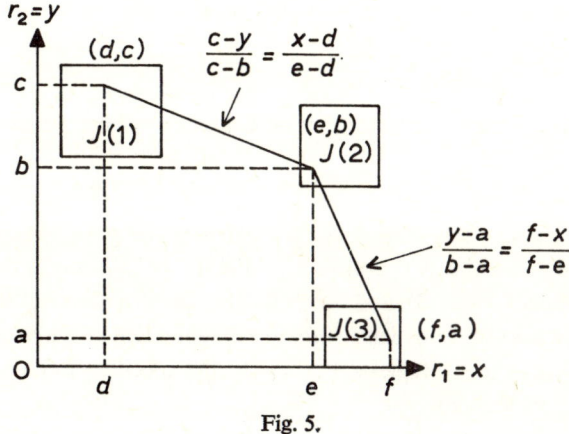

Fig. 5.

timal. The proof that $r_2(x)$, $x = 1, 3$ is semioptimal follows an analogous argument and will not be repeated. (Refer to Figure 5.) Let $(d, e, f) = (r_1(1), r_1(2), r_1(3))$ and $(c, b, a) = (r_2(1), r_2(2), r_2(3))$. Observe that $a < b < c$ and $d < e < f$, because $u_1(3) > u_1(2) > u_1(1)$ and $u_2(1) > u_2(2) > u_2(3)$.

Since the compromise solution in the reported utility space is Pareto optimal, according to our assumptions it should be either the segment $[(f, a), (e, b)]$ or the segment $[(e, b), (d, c)]$.[8]

Case 1. Suppose the compromise solution is on segment $[(f, a), (e, b)]$. To simplify our presentation, we replace (r_1, r_2) by (x, y). The segment can be written

(13) $$\frac{y-a}{b-a} = \frac{f-x}{f-e} \qquad e \leq x \leq f.$$

The compromise solution (x_p, y_p) in the reported utility space must minimize

(14) $$D_p(x, y) = (f - x)^p + (c - y)^p$$

Put (13) in (14) to get

(15) $$\bar{D}_p(y) = \left(\frac{f-e}{b-a}\right)^p (y - a)^p + (c - y)^p.$$

By setting the derivative of $\bar{D}_p(y)$ equal to zero in (15), we get

$$y_p = \frac{c + \left(\frac{f-e}{b-a}\right)^{p/p-1} a}{1 + \left(\frac{f-e}{b-a}\right)^{p/p-1}} = a + \frac{c-a}{1 + \left(\frac{f-e}{b-a}\right)^{p/p-1}}.$$

Since $u_1(2) < u_1(3)$, the smaller is y_p, the better the payoff is to Player 1. Note, e and f are the two controls of Player 1. The minimum of y_p is achieved when $f = \bar{I}(1, 3)$ and $e = \underline{I}(1, 2)$. Observe that y_p is independent of d and that the above decisions are independent of the other decision variables. Thus the strategies $r_1(x)$, $x = 1, 2, 3$ specified in (i) and (ii) are semioptimal in this case.

Case 2. Suppose the compromise solution is on the segment $[(e, b), (d, c)]$ which can be represented by

(16) $\quad \dfrac{c-y}{c-b} = \dfrac{x-d}{e-d}.$

The compromise solution (x_p, y_p) must minimize $D_p(x, y)$ defined by (14). Putting (16) in (14), we get

$$\bar{D}_p(x) = (f-x)^p + \left(\frac{c-b}{e-d}\right)^p (x-d)^p.$$

Setting the derivative of $\bar{D}_p(x)$ equal to zero, we obtain

(17) $\quad x_p = \dfrac{f + \left(\frac{c-b}{e-d}\right)^{p/p-1} d}{1 + \left(\frac{c-b}{e-d}\right)^{p/p-1}}.$

Since $u_1(1) < u_1(2)$, the smaller is y_p, the better the payoff is to Player 1. Observe that minimizing y_p is equivalent to maximizing

(18) $\quad g(e, d, f) = \dfrac{c - y_p}{c - b} = \dfrac{x_p - d}{e - d} = \dfrac{f - d}{e - d} \dfrac{1}{1 + \left(\frac{c-b}{e-d}\right)^{p/p-1}}.$

The second and the third equalities come from (16) and (17) respectively.

Observe that $f = I(1, 3)$ maximizes $g(e, d, f)$ independently of the other decision variables of the two players. Thus $r_1(3) = I(1, 3)$ is a semioptimal control in this case.

Next we shall show that $r_1(1) = I(1, 1)$ is also a semioptimal control. Toward this end, we rewrite (18) into

$$(19) \quad g(e, d, f) = \frac{f - e + e - d}{(e - d) + \dfrac{(c - b)^{p/p-1}}{(e - d)^{1/p-1}}} =$$

$$= \frac{(f - e)(e - d)^{1/p-1} + (e - d)^{p/p-1}}{(e - d)^{p/p-1} + (c - b)^{p/p-1}} =$$

$$= 1 + \frac{(f - e)(e - d)^{1/p-1} - (c - b)^{p/p-1}}{(e - d)^{p/p-1} + (c - b)^{p/p-1}}.$$

Let
$$A = f - e$$
$$B = (c - b)^{p/p-1}$$
$$X = e - d = X(d).$$

Then $g(e, d, f) = 1 + h(X)$ where

$$h(X) = \frac{AX^{1/p-1} - B}{X^{p/p-1} + B}.$$

Observe that $y_p \geq b$. Thus $g(e, d, f) = c - y_p/c - b \leq 1$ and $h(X) \leq 0$, which holds if $X^{1/p-1} \leq B/A$. On the other hand $X = e - d > 0$. Thus $0 < X^{1/p-1}$.

We want to study the sign of the derivative of $h(X)$ for X such that $0 < X^{1/p-1} \leq B/A$.

The numerator of $dh(X)/dX$ is given by

$$N(X) = \frac{1}{p-1} AX^{(2-p)/p-1}(X^{p/p-1} + B) +$$

$$- \frac{P}{p-1} X^{1/p-1}(AX^{1/p-1} - B) =$$

$$= X^{(2-p)/p-1} \left\{ -AX^{p/p-1} + \frac{P}{p-1} BX + \frac{AB}{p-1} \right\}.$$

By Descartes' Rule, the expression in braces has exactly one positive zero. But at $X = 0$ $N(X) = AB/(p-1) > 0$ and at $X^{1/p-1} = B/A$, $N = -A(B/A)X + P/(p-1) BX + AB/(p-1) > 0$. It follows that $N(X)$ is positive over X

such that $0 < X^{1/p-1} \leq B/A$. Thus $h(X)$ is an increasing function of X in that interval, or $g(e, d, f)$ is a decreasing function of d. In order to maximize $g(e, d, f)$, we should put $r_1(1) = d = I(1, 1)$. Again the choice of $r_1(1) = I(1, 1)$ is independent of the choice of other variables. Thus $r_1(1) = I(1, 1)$ is also a semioptimal control in this case.

It remains to show that if $I(1, 2) + I(2, 2) > \max\{I(1, 3) + I(2, 3), I(2, 1) + I(1, 1)\}$ and $p \geq 2$, then μ_1 defined in (ii) is a semioptimal strategy. In order to achieve this we rewrite (18) into

$$(21) \quad g(e, d, f) = \frac{(f - d)(e - d)^{1/p-1}}{(e - d)^{p/p-1} + (c - b)^{p/p-1}} = \frac{DY^{1/p-1}}{Y^{p/p-1} + B} \equiv k(Y),$$

where
$$D = f - d$$
$$Y = e - d = Y(e)$$
$$B = (c - b)^{p/p-1}$$

$$k'(Y) = \frac{D}{(Y^{p/p-1} + B)^2} Y^{(2-p)/p-1} \left[\frac{B}{p - 1} - Y^{p/p-1} \right].$$

From (21) and the definition of Y we have

$$\frac{\partial g(e, d, f)}{\partial e} = k'(Y) = \frac{D}{(Y^{p/p-1} + B)^2} Y^{(p/p-1)-2} \times$$
$$\times \left[\frac{(c - b)^{p/p-1}}{p - 1} - (e - d)^{p/p-1} \right].$$

Since $D > 0$ and $Y > 0$, we have

$$\frac{\partial g}{\partial e} < 0 \quad \text{if and only if} \quad \frac{(c - b)^{p/p-1}}{p - 1} <$$
$$< (e - d)^{p/p-1} \quad \text{or} \quad \left(\frac{c - b}{e - d} \right)^{p/p-1} < p - 1.$$

This will hold for all $p \geq 2$ provided that $(c - b)/(e - d) < 1$ or, equivalently, $c + d < b + e$. But we have already established that $d = r_1(1) = I(1, 1)$ is a semioptimal control. Furthermore, $c = r_2(1) \leq I(2, 1)$, $b = r_2(2) \geq I(2, 2)$ and $e = r_1(1) \geq I(1, 2)$. Hence $c + d \leq I(2, 1) + I(1, 1) < I(2, 2) + I(I, 2) \leq \leq b + e$ (the middle inequality holding by virtue of the above assumption). Thus we see that $\partial g/\partial e$ is negative whenever our assumptions are satisfied.

In order to maximize g, we should put $e = r_1(2) = I(1, 2)$. This time the control is not independent of all other variables, but does not depend on the other player's choices. Thus the strategy $\mu_1 = (I(1, 1), I(1, 2), I(1, 3))$ is semioptimal.

Q.E.D.

Proof for (iii). From the assumption, $u(2)$ lies above the line $[u(1), u(3)]$. The payoffs resulting from the compromise solutions all lie in one of the segments $[u(1), u(2)]$ or $[u(2), u(3)]$. Thus they are Pareto optimal, and the strategies we specified are optimal.

Q.E.D.

REMARK 3.3.3. Observe that if $J(2)$ does not lie on top of the line $[A_3, B_3]$ the resulting payoff of the strategies specified in the theorem may be not Pareto optimal and thus the strategies are not optimal. In order to see this point, let us consider the example depicted in Figure 6.

Observe that according to our compromise scheme, the choice 2 will be selected for $p \geq 1$. Also observe that the payoff (the true utility) is represented by the point $u(2)$ which lies below the line of $[u(1), u(3)]$ (or the line $[A_3, B_3]$ in our previous context). Clearly the payoff is not a Pareto optimal solution.

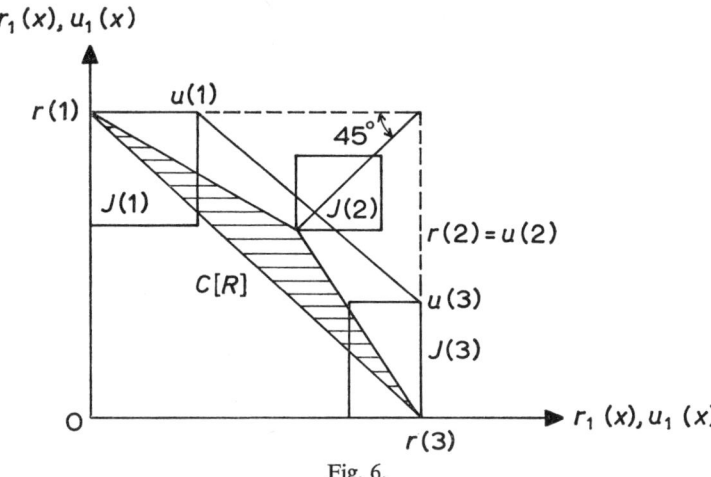

Fig. 6.

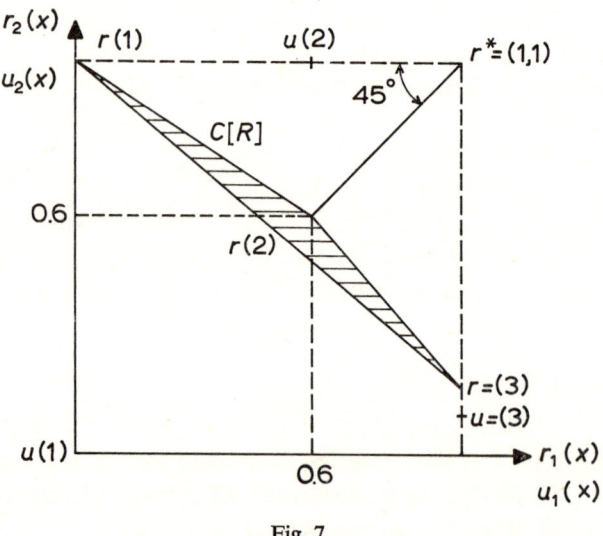

Fig. 7.

REMARK 3.3.4. The no cheating assumption stated at the beginning of this section cannot be eliminated. As an example, let $u(1)=(0,0)$, $u(2)=(0.6, 1)$, $u(3)=(1, 0.1)$, $I(1, 1)=\{0\}$, $I(1, 2)=\{0.6\}$, $I(1, 3)=\{1\}$, (i.e., the true utility of player 1 is known) and $I(2, x) = [0, 1]$ $x=1, 2, 3$. (See Figure 7)

Then an optimal strategy for player 2 is

$$r_2(1) = 1$$
$$r_2(2) = 0.6$$

and

$$0 \leqq r_2(3) < 0.2.$$

This is a great cheating strategy which is not permitted in the theorem.

4. CONCLUSION

We have investigated the impact of the compromise parameter p and the nature of the reporting games in the compromise scheme. Our results show a nice connection between compromise solutions and simple majority rule. Through the concept of semioptimality and optimality we derive some readily applicable results in some classes of reporting games. Nevertheless, our results are still not the most general ones. The reporting games with n-person and q choices $n \geq 2$, $q > 3$ still remain to be explored. Some necessary and sufficient conditions for a game to have semioptimal and optimal strategies are currently under study.

5. Acknowledgement

The authors thank Professor A. Rapoport for his helpful comments on [12] and his encouragement of this research.

6. Appendix

Proof of Lemma 3.1.5. Given $p > 1$ and a feasible report $R = (r(1), ..., r(q))$, let[9] $r^0 = \sum_x \lambda(x) r(x)$ be the compromise solution with respect to p in the reported utility space $C[R]$. Now let $\{S^k\}$, $S^k = (s^k(1), ..., s^k(q))$, be a sequence of feasible reports such that $\{S^k\}$ converges to R. Let $s^k = \sum_x \mu^k(x) s^k(x)$ be the compromise solution in the reported utility space $C[S^k]$ with parameter p. In view of (2), to show that $P_j(R; p)$ is continuous in R, it suffices to show that $\{\mu^k(x)\} \to \lambda(x)$, for all $x \in X$.

Our first goal is to establish

(A) $\quad \sum_x \mu^k(x) r(x) \to \sum_x \lambda(x) r(x) = r^0 \quad \text{as} \quad k \to \infty.$

Observe that for any feasible report $S = \{s(x) \mid x \in X\}$, if $|s_j(x) - r_j(x)| < \varepsilon$ all $j = 1, ..., n$ and $x = 1, ..., q$, then

(B) $\quad |s_j^* - r_j^*| < \varepsilon$ all j
$\|r^* - s^*\|_p < [n\varepsilon^p]^{1/p} = n^{1/p}\varepsilon$, ($r^*$, s^* are reported utopia points) and
$\|r(x) - s(x)\|_p < n^{1/p}\varepsilon$, all x.

Now let $s^0 = \sum_x \mu(x) s(x)$ be the compromise solution with respect to $C[S]$. then

(C) $\quad \|s^* - \sum_x \mu(x) s(x)\|_p \leq \|s^* - \sum_x \lambda(x) s(x)\|_p \leq$
$\leq \|s^* - r^*\|_p + \|r^* - \sum_x \lambda(x) r(x)\|_p +$

$$+ \sum_x \lambda(x) \|r(x) - s(x)\|_p <$$
$$< \|r^* - \sum_x \lambda(x) r(x)\|_p + 2n^{1/p}\varepsilon.$$

The first inequality comes from the definition of compromise solution; the second inequality comes from triangle inequalities and the last inequality comes from (B) and $\sum_x \lambda(x) = 1$. Thus

(D) $\quad \|r^* - \sum_x \mu(x) r(x)\|_p \leq \|r^* - s^*\|_p + \|s^* - \sum_x \mu(x) s(x)\|_p +$
$$+ \sum_x \mu(x) \|s(x) - r(x)\|_p <$$
$$< \|s^* - \sum_x \mu(x) s(x)\|_p + 2n^{1/p}\varepsilon <$$
$$< \|r^* - \sum_x \lambda(x) r(x)\|_p + 4n^{1/p}\varepsilon =$$
$$= \|r^* - r^0\|_p + 4n^{1/p}\varepsilon.$$

The first inequality comes from triangle inequalities, the second inequality comes from (B) and $\sum_x \mu(x) = 1$, the last inequality comes from (C).

Replace S and $\mu(x)$ respectively by S^k and $\mu^k(x)$ and ε by $\varepsilon^k = \max_{j,x} |s_j^k(x) - r_j(x)|$. Observe that $\varepsilon^k \to 0$ as $k \to \infty$. From (D) we get

(E) $\quad \|r^* - \sum_x \mu^k(x) r(x)\|_p \leq \|r^* - r^0\|_p + 4n^{1/p}\varepsilon^k.$

Let r' be any limit point of the sequence $\sum_x \mu^k(x) r(x)$ as $k \to \infty$.
Then
$$\|r^* - r'\|_p = \|r^* - r^0\|_p \quad \text{(by (E))}.$$

Since r^0 is the unique compromise solution with respect to $C[R]$ when $p > 1$, we see that $r' = r^0$. Thus (A) is established. Now let $\{v(x) | x \in X\}$ be any limit point of the sequence $\{u^k(x) | x \in X\}$.
Then
$$\sum_x \mu^k(x) r(x) \to \sum_x v(x) r(x) = \sum_x \lambda(x) r(x) = r^0.$$

So by the uniqueness of the representation of r^0, $v(x) = \lambda(x)$ for all $x \in X$, Hence $\mu^k(x) \to \lambda(x)$ for all $x \in X$ as we want to establish.

Q.E.D.

University of Rochester, New York

NOTES

[1] By $u(x) \geq 0$, we mean that $u_j(x) \geq 0$ for all $j \geq 1, \ldots, n$.
[2] The convexity on U_0 could be relaxed to $\Lambda \leq$-convexity. For cone convexity see [15].
[3] When u (1) is not unique, we consider its least upper bound or greatest lower bound. The result still holds.
[4] Thus, $u_1 + g(u_1)$ is concave.
[5] Despite this terminology, we will consider games in which the set $I(j, x)$ is disconnected (not an interval).
[6] This can only happen when one of the k-dimensional faces of the Pareto optimal set in $C[R]$ contains more than $k+1$ points, which might well be considered an event of 'probability zero.' (by random selection it is very difficult to have more than $k+1$ points lying on k-dimensional faces; without premeditation, in reporting games this hardly can occur).
[7] If the set of optimal strategies is not unique, then by varying the set of optimal strategies, the payoffs to the individual players may be not unique, and thus the values of the game is not defined.
[8] If the point (e, b) lies below the line joining (d, c) and (f, a) then Theorem 3.2.1 can be applied as in the proof of Theorem 3.3.1.
[9] We slightly simplify our notation to facilitate our presentation in this proof.

BIBLIOGRAPHY

[1] Arrow, K. J., *Social Choice and Individual Values*, Cowles Commission Monograph 12, 1951.
[2] Blaquiere, A., Gerard, F., and Leitmann, G., *Quantitative and Qualitative Games*, Academic Press, 1969.
[3] Ho, Y. C., *Final Report of the First International Conference on the Theory and Applications of Differential Games*, Amherst, Massachusetts, January, 1970.
[4] Isaacs, R., *Differential Games*, John Wiley, 1965.
[5] Luce, R. D., and Raiffa, H., *Games and Decisions*, John Wiley and Sons, Seventh printing, 1967.
[6] Olson, M. Jr., *The Logic of Collective Action – Public Goods and the Theory of Groups*, Harvard University Press, 1965.
[7] Owen, G., *Games Theory*, Saunders Company, 1968.
[8] Rapoport, A., *n-Person Game Theory – Concepts and Applications*, The University of Michigan Press, 1970.
[9] Shapley, L. S., 'A Value for n-Person Games', Contributions to the Theory of Games II, (ed. by H. W. Kulm and A. W. Tucker), *Annals of Math. Studies* **28**, (1953) 307–317.
[10] Shubik, M., *Readings in Game Theory and Political Behavior*, Doubleday, 1954.
[11] Von Neumann, J. and Morgenstern, O., *Theory of Games and Economic Behavior*, Princeton Press, 1947.
[12] Yu, P. L., *A Class of Solutions for Group Decision Problems*, Center for System Science, CSS 71-06, University of Rochester, 1971. (To appear in Management Science.)
[13] Rapoport, A., Private Communication.
[14] Freimer, M. and Yu, P. L., 'Some New Results on Compromise Solutions',

(forthcoming), Systems Analysis Program Working Paper Series No. F7221, Graduate School of Management, University of Rochester, Rochester, New York.
[15] Yu, P. L., *Cone Convexity, Cone Extreme Points and Nondominated Solutions in Decision Problems with Multiobjectives*, Center for System Science, CSS 70-02, Graduate School of Management, University of Rochester, February, 1972. (To appear in *Journal of Optimization Theory and Application*, in two parts: Part I: *Cone Convexity and Cone Extreme Points*, Part II *Nondominated Solutions*.)
[16] Glicksberg, I. L., 'A Further Generalization of the Kakutani Fixed Point Theorem with Application to Nash Equilibrium Points', *Proceedings of the American Mathematical Society*, Vol. III, 1952.

NIGEL HOWARD

'GENERAL' METAGAMES: AN EXTENSION OF THE METAGAME CONCEPT

I. INTRODUCTION

If G is a game in normal form, and if k is a player in G, the (first-level) metagame kG, as defined in [2], is the normal-form game that would exist if player k chose his strategy in G in knowledge of the other players' strategies (in G).

Hence, by recursion, the second-level metagame jkG, where j and k are players, is the game in which j chooses his strategy (in kG) in knowledge of the other's strategies (in kG); in terms of strategies in G, it is a game in which j reacts (a) to k's reactions to the actions of the players other than k; (b) to the actions of the players other than j and k. Continuing in this way, we form the set of all metagames $k_1 \ldots k_r G$, where each k_i is a player, and r is an integer $\geqslant 0$. This set of metagames is called the *metagame tree*.

In [2] these metagames are investigated, not because it is supposed that players will somehow have this kind of knowledge (to suppose that would be to contradict the assumptions underlying the normal form G) but because it is observed that normal-form players frequently attempt to predict (without, of course, any guarantee of success) what strategies each of the other players will choose. They do this, naturally, in order to plan their own strategies; and this leads us to consider an attractive definition of 'stability'. An outcome (in the sense of a strategy n-tuple) is defined as 'stable' precisely when all players do in fact succeed in predicting it. This is an important kind of stability, worth game-theoretic study. It is clear, however, that in order to study it the kind of complexities found in higher-level metagames need to be considered.

Our reason, then, for investigating metagames is to investigate this condition of 'stability' – i.e. the condition of mutual prediction by the players of each others' strategies. But for this purpose the definition of a metagame used in [2] is not wide enough. For this purpose what we really need is a definition of 'metagame' that is wide enough to cover all possible kinds of mutual prediction by players of one anothers' strategies. That this is by no means achieved by the definition of a metagame used in [2] will become apparent as we proceed.

In this paper I shall generalize the metagame concept so that (as far as I can see) it does cover all possible kinds of mutual prediction. I shall then reestablish for these 'general' metagames certain properties proved in [2] for the 'special' metagames considered there. I shall also show how in some respects the set of general metagames has a more appealing structure than the set of 'special' ones.

Among the properties of metagames proved in [2] are those by virtue of which metagame theory can be said to *unify* the somewhat separate cooperative and non-cooperative approaches used in game theory by providing a single framework within which both cooperative and non-cooperative solution concepts, usually put forward as separate definitions, are obtained from theorems characterizing the properties of the metagame tree. These fundamental properties, discussed below in Section III, will be among those re-proved for the case of 'general' metagames.

II. PRELIMINARIES

We write $G=(S_i, M_i: i\in N)$ for a game in normal form. Here N is the set of players and S_i is player i's strategy set. Both N and S_i are assumed to have at least two elements. We write s_i for a typical element of S_i; and in general, $s_K = (s_i: i\in K)$ will represent, for any $0 \subset K \subseteq N$, a typical element of the set $S_K = \prod_{i\in K} S_i$. Thus s_K is a typical strategy for the coalition K (where a coalition is any non-empty subset of players). S_N is also written S, with typical element $s=(s_i: i\in N)$, and is called the set of outcomes. Next, for each player i, M_i is player i's preference function. This is a set-valued function $M_i: S \to B(S)$ (where $B(S)$ is the set of all subsets of S). This set-valued function represents i's preference as follows: '$s \in M_i \bar{s}$' (where $s, \bar{s} \in S$) is read 'S is not preferred by i to $\bar{s}$'. This is clearly a general form of preference function that covers the case of numerical utilities as well as other possibilities. We assume only that $s \in M_i s$ for all $s \in S$. This again is rather general – preferences need not be transitive, for example.

Notice that we write '0' for the empty set and '$\subset$' for the relation 'is a *proper* subset of', reserving '$\subseteq$' for 'is a subset of'. Hence, above we have defined S_K only for $K \neq 0$. If $K=0$, let us define $S_K = 0$. Hence, in this case

s_K (a typical element of S_K) represents nothing, and may be regarded as equivalent to a blank space on the page.

We shall partition s freely, allowing ourselves to write $s = (s_K, s_{N-K})$ for any $K \subseteq N$. For $K \neq 0$ we define *coalition preference functions* M_K as follows

$$\forall s: M_K s = \bigcup_{i \in K} M_i s.$$

Hence, $s \in M_K \bar{s}$ remains true for coalitions as for individual players. We also use the functions $\tilde{M}_i$, $\tilde{M}_K$, defined by $\tilde{M}_i s \equiv S - M_i s$, $\tilde{M}_K s \equiv S - M_K s$. One sees that '$s \in \tilde{M}_i \bar{s}$' means '$i$ prefers s to $\bar{s}$', while '$s \in \tilde{M}_K \bar{s}$' mean "all members of K prefer s to $\bar{s}$'.

The normal-form game kG is derived from G by replacing the set S_k with the set $S_k^{S_{N-k}}$ of all functions from S_{N-K} to S_k. (As here, we shall often write k for $\{k\}$ when no confusion is possible.) Thus

$$kG = (S'_i, M'_i : i \in N),$$

where $S'_i = S_i$ for $i \in N - k$, while $S'_i = S_k^{S_{N-k}}$. The functions $M'_i : S_{N-k} \times \times S_k^{S_{N-k}} \to \beta(S_{N-k} \times S_k^{S_{N-k}})$ are defined as follows. Write (s_{N-k}, f) for a typical outcome of kG. Write $\beta(s_{N-k}, f) = (s_{N-k}, f(s_{N-k}))$ for the corresponding outcome of G. Thus β is an operator that maps outcomes of a metagame to the corresponding outcomes of the game from which the metagame is derived. Then M'_i is defined by writing

$$x \in M'_i \bar{x} \Leftrightarrow \beta x \in M_i \beta \bar{x},$$

where x, $\bar{x}$ are outcomes of kG.

Above we have defined the normal form of kG. But kG may instead be regarded as a two-stage game in extensive form as follows. *Stage I.* Each $i \in N - k$ chooses a strategy s_i. *Stage II.* Player k, being informed of the choices made at Stage I, chooses a strategy s_k.

As an example, let $N = \{1, 2\}$ and let $S_1 = S_2 = \{a, b, c\}$. Then in the game $2G$, player 2 has 27 strategies consisting of the functions $f: \{a, b, c\} \to \{a, b, c\}$. One such strategy is the $f = \hat{f}$ defined by $\hat{f}(x) = x$ for all $x \in \{a, b, c\}$. This $\hat{f}$ might be called 'tit-for-tat'.

Other metagames are derived by recursive application of the process by which kG is derived from G. The general recursive step is as follows. Let $H = k_1 \ldots k_r G$ be an rth-level metagame. Write $X(H) = \prod_{i \in N} X_i(H)$ for the set of outcomes of H, so that we can write $H = (X_i(H), M'_i : i \in N)$. The $(r+1)$th-level metagame $kH = kk_1 \ldots k_r G$ is now constructed by substituting $X_k(H)^{X_{N-\kappa(K)}}$ for $X_k(H)$ and M_i'' for each M_i', where M_i'' is such that for all $x, \bar{x} \in X(kH)$:

$$x \in M_i'' \bar{x} \Leftrightarrow \beta x \in M_i' \beta \bar{x}.$$

By following through this recursion it can readily be seen that M_i'' is given, in terms of the basic preferences M_i, by

$$x \in M_i'' \bar{x} \Leftrightarrow \beta^* x \in M_i \beta^* \bar{x}$$

– where 'β^*' stands for a sufficient number of applications of the β operator (in this case $r+1$ applications, so that here β^* stands for β^{r+1}) to map any metagame outcome into the unique corresponding basic outcome.

III. THE METAGAME DERIVATION OF 'COOPERATIVE' AND 'NON-COOPERATIVE' CONCEPTS

At the basis of 'non-cooperative' game theory is the concept of an *equilibrium point*. But a more fundamental concept underlying this *is that of a rational outcome for player i* (or *i-rational outcome*), defined as an $\bar{s}$ obeying the condition $\forall s_i : (s_{N-i}) \in M_i \bar{s}$. Writing $R_i(G)$ for the set of all *i*-rational outcomes, we form the usual set $E(G)$ of equilibrium points by writing $E(G) = \bigcap_{i \in N} R_i(G)$.

A natural generalization of $R_i(G)$ is to define $R_K(G)$, where K is a coalition, as the set of all $\bar{s}$ obeying

(1) $\quad \forall s_K : (s_K, \bar{s}_{N-K}) \in M_K \bar{s}.$

We call such an $\bar{s}$ a *K-rational outcome*. The concepts of a metagame and of a rational outcome are the two 'primitive' concepts we use, and both are directly connected with the definition of stability discussed above. This connection, in the case of the metagame concept, has already been discussed. In regard to rational outcomes, the connection is that if a player *i* is 'rational' in the strict sense of optimizing with respect to his preferences, one can deduce immediately that he will never predict an outcome that is not *i*-rational.

We have seen that the *equilibrium points* of a game are immediately derivable from the sets of rational outcomes. So, too, are Aumann's [1] *strong equilibria*. These are the outcomes in the set $\bigcap_{0 \subset K \subseteq N} R_K(G)$. Next, consider the outcomes $\bar{s}$ that obey the condition:

(2) $\quad \forall s_K \exists s_{N-K}: s \in M_K \bar{s}.$

These are just the outcomes that correspond to payoff vectors that Aumann [1] calls 'α-undominated through K'. They are, therefore, important in so-called "cooperative' game theory. Equally important are the $\bar{s}$ obeying the condition

(3) $\quad \exists s_{N-K} \forall s_K: s \in M_K \bar{s}.$

These are the outcomes that, according to Aumann, have payoff vectors 'β-undominated through K'.

In reference [2], conditions (2) and (3) are obtained as characterizations of certain important properties of the metagame tree. The development is as follows. For any game or metagame H, let $\mathscr{D}(H)$ be the set of metagame 'descendants' of H – i.e. the set of metagames of the form $k_1 \ldots k_r H$ (where $r \geqslant 0$). Then for any given basic game G and any coalition K one is naturally interested in defining and characterizing the following sets:

$$\bigcap_{H \in \mathscr{D}(G)} \beta^* R_K(H); \quad \bigcup_{H \in \mathscr{D}(G)} \beta^* R_K(H).$$

In [2], the first set is proved equal to $R_K(G)$ itself. Hence, $R_K(G)$ is not only the set of K-rational outcomes of G, it is also the set of outcomes that are K-metarational from every metagame (where $\bar{s}$ is called 'K-metarational from H' if it belongs to $\beta^* R_K(H)$). Next, the second set is denoted $\Gamma_K(G)$, is called the *set of general K-metarational* outcomes, and is proved equal to the set of all $\bar{s}$ obeying condition (2). Hence, condition (2) characterizes just those $\bar{s}$ that are *K-metarational from some metagame*. Finally, one forms the set

$$\bigcap_{H \in \mathscr{D}(G)} \beta^* \Gamma_K(H).$$

This is the set of outcomes that are *K-metarational from some descendant of every metagame*. In [2] it is denoted $\Sigma_K(G)$ and proved equal to the set of $\bar{s}$ obeying (3).

It follows from the above that the set of outcomes yielding payoffs in Aumann's *α-core* is the set $\bigcap_{0 \subset K \subseteq N} \Gamma_K(G)$, while those with payoffs in the *β-core* is $\bigcap_{0 \subset K \subseteq N} \Sigma_K(G)$.

IV. GENERALIZED METAGAMES

In the metagame kG, player k succeeds in predicting the others' strategies exactly. There is no metagame in which he achieves only partial prediction. Also, he is the only player who (at this stage) has any power to predict; the other players do not predict at all.

A generalized metagame in which these deficiencies are overcome may be defined as follows. Let P_i, for each $i \in N$, be a given *partition* of S_i (i.e. a set of non-empty, disjoint sets such that $\bigcup P_i = S_i$) and let $P = (P_i : i \in N)$.

		Player 2 (Soviet Union)		
		a Disarm, Allow Inspection	*b* Disarm, Not Allow Inspection	*c* Not Disarm, Not Allow Inspection
Player 1 (U.S.)	*a* Disarm, Allow Inspection	6, 5	4, 7	1, 9
	b Disarm, Not Allow Inspection	7, 4	5, 6	2, 8
	c Not Disarm, Not Allow Inspection	9, 1	8, 2	3, 3

Fig. 1. Disarmament Game with Inspection. Figures in cells represent player's ordinal preferences – e.g. (c, a) is the most preferred outcome for the U.S., while (a, c) is least preferred.

Then define PG as a game played (in extensive form) as follows. *Stage I.* Each player $i \in N$ chooses a p_i from the given set P_i. *Stage II.* Each player $i \in N$, having been informed of all the choices made at Stage I, chooses a strategy s_i from the set p_i chosen by him in Stage I.

The effect of this is to generalize the concept of a metagame in precisely the desired manner, since under this scheme i's choice of s_i may be partially, yet not wholly, predictable by the other players. It will be wholly predictable only if the p_i that contains s_i is a unit set. At the other extreme, if $P_i = \{p_i\} = \{S_i\}$, then s_i is entirely unpredictable. Also, because partial prediction is possible, it becomes possible for player i's choice of an s_i to be both based on a (partial) prediction and itself (partially) predictable.

To illustrate consider as before a game in which $N = \{1, 2\}$ and $S_1 = S_2 = \{a, b, c\}$. Suppose the (ordinal) preferences of the players are as indicated in Figure 1. (Recall that ordinality is not assumed in general.) In Figure 1 this game has been interpreted as a disarmament problem involving inspection. This gives motivation for investigating the metagame $\bar{P}G$ in which $\bar{P}_1 = \bar{P}_2 = \{\{a\}, \{b, c\}\}$, since this metagame models the assumption that, without inspection, neither player will know whether the other has disarmed. The lines drawn in the matrix in Figure 1 indicate the partitions $\bar{P}_1, \bar{P}_2$. We may now imagine that $\bar{P}G$ is played by each player first separately choosing a section of his side of the matrix, thus determining which of four possible subgames is then to be played in 'one-shot' fashion.

I have indicated that each indexed set $P = (P_i : i \in N)$, where each P_i is a partition of S_i, determines one of these generalized metagames. The special metagame kG can be seen to be equivalent to the generalized metagame PG in which $P_k = \{S_k\}$ and, for each $i \in N - k$, $P_i = \{\{s_i\} \mid s_i \in S_i\}$; for in this metagame, as in kG, player k can predict exactly what strategy each other player in $N - k$ will choose while they can predict nothing about his choice. So far, however, an unnecessary restriction has been imposed on $P = = (P_i : i \in N)$. I have assumed that any strategy of a particular player i belongs to a unique $p_i \in P_i$. Hence, if i decides to choose the strategy s_i, he cannot further influence the prediction $p_i \ni s_i$ that the other players will arrive at. However, real life predictions are often arrived at by players informing each other of the strategy they intend to choose (though nothing prevents them from lying to one another); and this, as well as other ways of arriving at predictions, will often enable i to exercise some influence over the prediction arrived at concerning s_i. So let us drop the restriction that each P_i be a *partition* of S_i. Let each P_i be any *cover* of S_i (i.e. any set of non-empty sets, not necessarily disjoint, such that $\bigcup P_i = S_i$). To illustrate, suppose that in the game of Figure 1 the U.S. and the U.S.S.R. may each choose, as an alternative to the choice of $\{a\}$

	{a }	{a, b}		{a, b, c}		
{a}	6, 5	4, 7	1, 9	6, 5	4, 7	1, 9
{b, c}	7, 4 9, 1	5, 6 8, 2	2, 8 3, 3	7, 4 9, 1	5, 6 8, 2	2, 8 3, 3
{a, b, c}	⑥, ⑤ 7, 4 9, 1	4, 7 5, 6 ⑧, ②	1, 9 2, 8 ③, ③	6, 5 7, 4 ⑨, ①	4, 7 5, 6 ⑧, ②	1, 9 2, 8 ③, ③

Fig. 2. Representation of the revised metagame $\bar{P}G$. The circled outcomes show a possible strategy for player 1.

or {b, c}, to reveal nothing at all about their intended strategies. Then $\bar{P}_1 = \bar{P}_2 = \{\{a\}, \{b, c\}, \{a, b, c\}\}$, and Figure 2 shows how $\bar{P}G$ must now be drawn if we wish to see it as a game in which players first choose sections of their sides of the matrix, then play a 'one-shot' subgame. A player's total strategy in $\bar{P}G$ can be shown by showing the (basic) strategy he decides conditionally to choose in each of the subgames within his chosen section; for example, Figure 2 shows the strategy for player 1 that entails choosing {a, b, c} in Stage I and, then, in Stage II, choosing strategy a if 2 chose {a} in Stage I and choosing strategy c otherwise.

Allowing $P = (P_i : i \in N)$ to be any indexed set such that each P_i is a cover of S_i, we find that there are

$$\prod_{i \in N} \gamma(|S_i|)$$

first-level metagames PG based on our basic game G – where $|S_i|$ is the number of elements in the set S_i and $\gamma(n)$ is the number of covers of a set with n elements. A recursive formula for $\gamma(n)$ is given by:

$$\gamma(1) = 1;$$
$$\gamma(n) = 2^{(2^n-1)} - 1 - \sum_{r=1}^{n-1} \binom{n}{r} \gamma(r),$$

so that $\gamma(3) = 109$, and the game in Figure 1 has 11, 181 metagames PG based on it. This compares with 2 special metagames kG and 25 metagames PG in which the covers P_i are restricted to be partitions.

The normal form of a general PG is as follows. Player i's strategy set

$X_i(PG) = \{(p_i, f_i) \mid p_i \in P_i; f_i \in p_i^{P_{N-i}}\}$, where $P_{N-i} = \prod_{j \in N-i} P_j$. This says that player i chooses: (i) a set $p_i \in P_i$ which may be termed his *commitment* (and which represents also the *prediction* that will be made by each other player); (ii) a function f_i from the set P_{N-i} of possible joint commitments by the other players to the set p_i representing his own commitment. This function f_i may be called his *policy*. For example, the circled outcomes in Figure 2 indicate the strategy $(\bar{p}_1, \bar{f}_1)$ for player 1 that is defined by

$$\bar{p}_1 = \{a, b, c\}; \quad \bar{f}_1\{a\} = a, \quad \bar{f}_1\{b, c\} = \bar{f}_1\{a, b, c\} = c.$$

An outcome of *PG* thus has the form $x = (p, f) = (p_i, f_i : i \in N)$; that is, it contains a commitment and a policy from each player. It gives rise, of course to the unique basic outcome

$$\beta x = \beta(p_i, f_i : i \in N) = (f_i(p_{N-i}) : i \in N) = \beta(p, f) = f(p),$$

– where $p_{N-i} = (p_j : j \in N - i) \in P_{N-i}$. Again we use '$\beta$' for an operator that takes any metagame outcome into the unique corresponding outcome of the game from which the metagame was derived. The notation $f(p)$ is used to represent the effect of the functions $f = (f_i : i \in N)$ on the arguments $(p_{N-i} : i \in N)$. Finally, then, the normal-form game *PG* is given by

$$PG = (X_i(PG), M'_i : i \in N)$$

where $X_i(PG)$ is as given above and M'_i is again given by

$$x \in M'_i \bar{x} \Leftrightarrow \beta x \in M_i \beta \bar{x}.$$

The $P = (P_i : i \in N)$ discussed above may be called a *prediction structure on G*. Having obtained the normal form of *PG*, where *P* is a prediction structure on *G*, we can by recursion obtain an infinite set of metagames of the form $P^{(r)} P^{(r-1)} \ldots P^{(1)} G$, where $r \geq 0$. All that is required is that each $P^{(k)}$ be a prediction structure on the game $P^{(k-1)} \ldots P^{(1)} G$. This infinite set of metagames based on *G* shall be denoted $\mathscr{C}(G)$. (Recall that $\mathscr{D}(G)$ denotes the set of 'special' metagames.)

The set $\mathscr{C}(g)$ can be widened further in the case of games with more than two players. We have assumed so far that the players other than i all form the same prediction p_i concerning s_i. This corresponds, perhaps, to a situation of open bargaining (or, perhaps, effective spying), wherein any communications from one player to another are overheard by all the players. We could model situations wherein different predictions may be formed by different players by defining a prediction structure on G as a $P = (P_i : i \in N)$ such that for each $i \in N$:

(i) P_i is a set with elements p_i;
(ii) each element $p_i \in P_i$ is an indexed set $p_i = (p_i^j : j \in N - i)$ of which each component p_i^j is a subset of S_i;
(iii) each $p_i \in P_i$ obeys $\bigcap p_i (= \bigcap_{j \in N - i} p_i^j) \neq 0$;
(iv) $\bigcup_{p_i \in P_i} \bigcap p_i = S_i$.

We could then define the extensive form of PG as follows. *Stage I.* Each player i chooses a vector p_i from the set P_i. *Stage II.* Each player j, having been informed of the jth component p_i^j of the vector p_i chosen by each player i in Stage I, chooses a strategy s_j from the set $\bigcap p_j$. The corresponding normal form is given by $PG = (X_i, M_i' : i \in N)$, where X_i is the set of all pairs (p_i, f_i) such that $p_i \in P_i$ and $f_i \in (\bigcap p_i)^{P^i}$ (where P^i is the Cartesian product $\prod_{j \in N - i} \{p_j^i \mid p_j \in P_j\}$ containing as elements all possible vectors $p^i = (p_j^i : j \in N - i)$ that i might be informed of when choosing in Stage II). The way this works is as follows. When each player i chooses a vector $p_i = (p_i^j : j \in N - i)$ of commitments that he chooses to make to each other player, then for each player j a vector $p^j = (p_i^j : i \in N - j)$ of predictions made by player j is determined; and the outcome $x \in X = \prod_{i \in N} X_i$, which may be written $x = (p_i, f_i : i \in N)$ to indicate the commitments made by each player, may equally well be written $x = (f_i, p^i : i \in N)$ to indicate the predictions they make. Writing it in this second form, we define the 'β' operator by

$$\beta x = \beta(f_i, p^i : i \in N) = (f_i(p^i) : i \in N);$$

and this enables us once again to define the preference functions M_i' by writing

$$x \in M_i' \bar{x} \Leftrightarrow \beta x \in M_i \beta \bar{x}.$$

A prediction structure P defined in this way will be called a *many-prediction* structure. The simpler case discussed previously, in which each P_i is a simple cover of S_i, will be called a *one-prediction* structure. Actually, in my work so far I have not found any use for many-prediction structures. One-prediction structures appear to be general enough for every kind of purpose, and I shall deal mostly with them. But for reference, form the infinite set of metagames of the form $P^{(r)} P^{(r-1)} \ldots P^{(1)} G$, where $r \geq 0$

and each $P^{(j)}$ is a many-prediction structure on $P^{(j-1)}\ldots P^{(1)}G$. Write $\bar{\mathscr{C}}(G)$ for this infinite set (recall that $\mathscr{C}(G)$ is the infinite set of one-prediction metagames with which we shall be principally concerned).

Returning to $\mathscr{C}(G)$, we have already seen that 'special' metagames are a special case. Certain other special cases are of interest. For any coalition K, let $P^{(K)} = (P_i^{(K)} : i \in N)$ be the structure on G given by:

(i) $P_i^{(K)} = \{S_i\}$, for all $i \in K$;
(ii) $P_i^{(K)} = \{\{s_i\} \mid s_i \in S_i\}$, for all $i \notin K$.

In $P^{(K)}G$, the players in the coalition K can predict exactly what strategies the others will choose, while they themselves are quite unpredictable. This game is a useful generalization of kG. Another special case is the prediction structure $\mathring{P} = (\mathring{P} : i \in N)$ such that $\mathring{P}_i = \{p_i \mid 0 \subset p_i \subseteq S_i\}$. This is the 'largest possible' structure, in the sense that if $P = (P_i : i \in N)$ is any one-prediction structure on G, then $\mathring{P}_i \supseteq P_i$ for all i. The game $\mathring{P}_i G$ is one in which any player can make any commitment to any other player. Notice that $P^{(K)}$ is a 'general' structure in that $P^{(K)}G$ is defined for any G that has the set N of players by merely stating: (i) for $i \in K$, $P_i^{(K)}$ is a unit set; (ii) for $i \notin K$, the elements of $P_i^{(K)}$ are unit sets. The structure $\mathring{P}$ is even more general, in that the definition I have given defines $\mathring{P}G$ for any game G.

Now let x be an outcome of a metagame $H \in \mathscr{C}(G)$ (or indeed, $\bar{\mathscr{C}}(G)$). By applying the 'β' operator to x a sufficient number of times, we obtain as before a unique outcome of G, denoted $\beta^* x$ as before. The set $B^* R_K(H)$ is called once again the set of $(K$-$H)$-*metarational outcomes* (or outcomes *metarational for K from H*, or K-*metarational* outcomes *from H*.) It will also be written $\hat{R}_K(H)$. All this is merely repeating, for the larger set $\mathscr{C}(G)$ or $\bar{\mathscr{C}}(G)$, what was said above for the set $\mathscr{D}(G)$ of special metagames. The set $\beta^* E(H)$, written $\hat{E}(H)$, is called the set of *metaequilibria* from H.

As an illustration, consider again the game $\bar{P}G$ shown in Figure 2, where $\bar{P}_1 = \bar{P}_2 = \{\{a\}, \{b, c\}, \{a, b, c\}\}$. In the basic game G we have $R_1(G) = \{(c, a), (c, b), (c, c)\}$ and $R_2(G) = \{(a, c), (b, c), (c, c)\}$. Hence $E(G) = \{(c, c)\}$. The strategy $(\bar{p}_1, \bar{f}_1)$ indicated in Figure 2 shows however, that $(a, a) \in \hat{R}_2$, since if player 1 chooses $(\bar{p}_1, \bar{f}_1)$ player 2 cannot do better than set $p_2 = \{a\}$. This (via the 'β' operator) leads to (a, a). We also have $(a, a) \in \hat{R}_1(\bar{P}G)$. But $(a, a) \notin \hat{E}(\bar{P}G)$. (Notice that $\hat{E}(\bar{P}G) = \beta^* [R_1(\bar{P}G) \cap R_2(\bar{P}G)]$, which is not the same as $\hat{R}_1(\bar{P}G) \cap \hat{R}_2(\bar{P}G) = \beta^* R_1(\bar{P}G) \cap$

$\cap \beta^* R_2(\bar{P}G)$.) Thus (a, a) is metarational for both players from $\bar{P}G$, but is not a metaequilibrium from this metagame.

In order to find a metagame from which (a, a) is a metaequilibrium, we have to go to a second-level metagame. Indeed, consider the metagame $\mathring{P}\bar{P}G$, where $\mathring{P}$, as before, is the prediction structure that allows each player to make any kind of commitment – meaning in this case any kind of 'second-level' commitment, that is, any subset of $X_i(\bar{P}G)$. We may visualize the game $\mathring{P}\bar{P}G$ being played as follows. *Stage I.* Each player i 'constrains' his own choice of (p_i, f_i) – i.e. his own choice of a strategy in $\bar{P}G$ – in any way he likes. *Stage II.* Each player i, having been informed of the 'constraints' chosen by the other player in Stage I, chooses a (p_i, f_i) from within the constraints he himself chose in Stage I. Here, a 'constraint' chosen by a player is a *statement* made by a player that constrains his choice of a (p_i, f_i). Thus, in this method of description, a second-level commitment chosen by a player is described, not by describing explicitly the set that constitutes the commitment, but by giving instead a statement that, indeed, defines a set because of the fact that a statement $M(x)$ about a set of objects with typical element x necessarily defines a set – i.e. the set $\{x \mid M(x)\}$. In this way, if P is a second-level structure, P_i is regarded as a set of *statements* that are called 'constraints'; and I find that this helps one to think about second-level metagames without getting one's levels confused.

We may denote by c_i the 'constraint' chosen by player i (c_i, therefore, has the form of a proposition about (p_i, f_i)). Thus, the following account defines a strategy for player 1 in the game $\mathring{P}\bar{P}G$. (i) Player 1 in Stage I chooses for c_1 the proposition '$f_1\{a\} = a$'. (ii) If c_2 logically implies the proposition '$f_2\{a\} = a$', player 1 in Stage II chooses (p_1, f_1) such that $p_1 = \{a\}$; but if c_2 does not logically imply this, he chooses (p_1, f_1) such that

$$p_1 = \{a, b, c\}, \quad f_1\{a\} = a, f_1\{b, c\} = f_1\{a, b, c\} = c.$$

A similar strategy may be defined for player 2 by substituting '1' for '2', and *vice versa*, in the foregoing account. When this is done, the reader may check that the two strategies are in equilibrium and lead (via two applications of the 'β' operator) to the basic outcome (a, a). Thus, $(a, a) \in \hat{E}(\mathring{P}\bar{P}G)$. On the other hand, $(b, b) \notin \hat{E}(\mathring{P}\bar{P}G)$: disarmament without inspection is unstable.

V. METARATIONAL OUTCOMES AND METAEQUILIBRIA

I shall now obtain characterizations of the following sets

$$\bigcap_{H \in \mathscr{C}(G)} \hat{R}_K(H);$$
$$\Gamma_K(G) = \bigcup_{H \in \mathscr{C}(G)} \hat{R}_K(H);$$
$$\Sigma_K(G) = \bigcap_{H \in \mathscr{C}(G)} \bigcup_{H' \in \mathscr{C}(H)} \hat{R}_K(H') = \bigcap_{H \in \mathscr{C}(G)} \hat{\Gamma}_K(H).$$

In the last line, $\hat{\Gamma}_K(H) = \beta^*\Gamma_K(H)$, where $\Gamma_K(H)$, for $H \in \mathscr{C}(G)$, is of course obtained from the definition of $\Gamma_K(G)$ by substituting H for the basic game G. Thus, $\Gamma_K(H) \subseteq X(H)$ (the set of outcomes of H), and the operator 'β' is needed to yield the corresponding set $\hat{\Gamma}_K(H)$ of 'basic' outcomes.

The above are sets of metarational outcomes. Their characterizations will be shown to be the same as the corresponding characterizations for the special metagame tree $\mathscr{D}(G)$; that is, identical characterizations are obtained if $\mathscr{D}(G)$ is substituted for $\mathscr{C}(G)$ in the above formulations. Hence, the characterizations we shall obtain, and the corresponding derivations of 'cooperative' and 'non-cooperative' game concepts, are the ones already described in Section III above.

I shall also characterize the sets defined as follows.

$$\bigcap_{H \in \mathscr{C}(G)} \hat{E}(H);$$
$$\Gamma(G) = \bigcup_{H \in \mathscr{C}(G)} \hat{E}(H);$$
$$\Sigma(G) = \bigcap_{H \in \mathscr{C}(G)} \bigcup_{H' \in \mathscr{C}(H)} \hat{E}(H') = \bigcap_{H \in \mathscr{C}(G)} \hat{\Gamma}(H).$$

Here the set $\hat{\Gamma}(H)$ is defined as equal to $\beta^*\Gamma(H)$ – c.f. the preceding remarks about $\hat{\Gamma}_K(H)$. $\Gamma(G)$ and $\Sigma(G)$ are respectively called the sets of *general* and *symmetric equilibria*. We shall prove that $\bigcap_{H \in \mathscr{C}(G)} \hat{E}(H)$ is equal to $E(G) (= \bigcap_{i \in N} R_i(G))$. This is the same as in the case of the special metagame tree $D(H)$. However, we also shall prove that $\Gamma(G) = \bigcap_{i \in N} \Gamma_i(G)$ – a proposition that does not hold for special metagames; that is, it does not hold when $\Gamma(G)$ is defined as equal to $\bigcup_{H \in \mathscr{D}(G)} \hat{E}(H)$. $\mathscr{C}(G)$ has a more appealing structure than $\mathscr{D}(G)$, and certain difficulties of interpretation encountered in reference [2] are completely overcome in the field of 'general' one-prediction metagames.

The above refers to the 'special' tree $\mathscr{D}(G)$ and the 'one-prediction' tree $\mathscr{C}(G)$. What can be said for the tree $\bar{\mathscr{C}}(G)$? Quite simply, all the foregoing statements remain true if $\bar{\mathscr{C}}(G)$ is substituted for $\mathscr{C}(G)$. By using the more complicated structures involved in the set $\bar{\mathscr{C}}(G)$, we change nothing.

To establish this, I shall first prove all theorems for the case of $\mathscr{C}(G)$, then state the 'meta-theorem' (which I shall leave to the reader to prove) that one can substitute $\bar{\mathscr{C}}(G)$ for $\mathscr{C}(G)$ throughout the proofs as well as throughout the theorems. Until further notice, then all metagames referred to are games in $\mathscr{C}(G)$.

To begin the mathematical development, if $x_K \in X_K(H)$ is a joint strategy for the coalition K in the metagame H, define $\langle x_K \rangle \subseteq X(H)$ as the set

$$\langle x_K \rangle = \{\bar{x} \in X(H) \mid \bar{x}_K = x_K\}.$$

Also, if $K=0$ define $\langle x_K \rangle$ as equal to the set $X(H)$ containing all the outcomes of H. The set $\langle x_K \rangle$ is thus the set of outcomes that are possible given that x_K has been chosen. The following lemmata are obtainable directly from this definition.

LEMMA I. If $K=N$, $\beta^*\langle x_K \rangle = \{\beta^* x_K\} =$ a unit set.
LEMMA II. If $K \supseteq J$, $\beta^*\langle x_K \rangle \subseteq \beta^*\langle x_J \rangle$.
LEMMA III. If $x = s \in S$, $\langle x_K \rangle \cap \langle x_{N-K} \rangle = \{s\} =$ a unit set.
LEMMA IV. $s \in \hat{R}_K(H) \Leftrightarrow \exists [x_{N-K} \in X_{N-K}(H)] : s \in \beta^*\langle x_{N-K} \rangle \subseteq M_K s$.
LEMMA V. $s \in \hat{E}(H) \Leftrightarrow \exists [x \in X(H)] \forall i : s \in \beta^*\langle x_{N-i} \rangle \subseteq M_i s$.

We now prove:

THEOREM I. If $\bar{s}$ is any basic outcome and P is any prediction structure on G, there exists $\bar{x} \in X(PG)$ such that:

$$\forall (0 \subset K \subseteq N) : \beta \langle \bar{x}_K \rangle = \langle \bar{s}_K \rangle.$$

Proof. Given $\bar{s}$, construct $\bar{x} = (\bar{p}, \bar{f}) = (\bar{p}_i, \bar{f}_i : i \in N)$ such that for any player i and any argument ξ of the function $\bar{f}_i$, we have $\bar{f}_i(\xi) = \bar{s}_i$ (that is, $\bar{f}_i$ is 'constant at' s_i). Then for any $s = (s_i : i \in N) \in \beta \langle \bar{x}_K \rangle$ we must have, for each $i \in K$, $s_i = \bar{s}_i$, since if $i \in K$, s_i is necessarily a value of $\bar{f}_i$. Hence, $\beta \langle \bar{x}_K \rangle \subseteq \langle \bar{s}_K \rangle$.

On the other hand, let $(\bar{s}_K, s'_{N-K})$ be an arbitrary element of $\langle \bar{s}_K \rangle$. We shall prove that it belongs to $\beta \langle \bar{x}_K \rangle$. The preceding paragraph establishes that for any $s' \in S$ there exists x' such that $\beta \langle x'_J \rangle \subseteq \langle s'_J \rangle$ for all $0 \subset J \subseteq N$. Hence, there exists x' such that $\beta \langle x'_{N-K} \rangle \subseteq \beta \langle s'_{N-K} \rangle$, and so in all we have

$$\beta \langle \bar{x}_K \rangle \cap \beta \langle x'_{N-K} \rangle \subseteq \langle \bar{s}_K \rangle \cap \langle s'_{N-K} \rangle.$$

Now from Lemma 2 we have

$$\beta\langle \bar{x}_K, x'_{N-K}\rangle \subseteq \beta\langle \bar{x}_K\rangle \cap \beta\langle x'_{N-K}\rangle$$

so that

$$\beta\langle \bar{x}_K, x'_{N-K}\rangle \subseteq \langle \bar{s}_K\rangle \cap \langle s'_{N-K}\rangle.$$

But from Lemma 1 the left-hand side of this is the unit set $\{\beta(\bar{x}_K, x'_{N-K})\}$, while from Lemma 3 the right-hand side is $\{(\bar{s}_K, s'_{N-K})\}$. Hence,

$$(\bar{s}_K, s'_{N-K}) = \beta(\bar{x}_K, x'_{N-K}) \in \beta\langle \bar{x}_K\rangle,$$

Q.E.D.

THEOREM II. *If H' is a descendant of H (i.e. if $H' \in \mathscr{C}(H)$) and $x \in X(H)$, there exists $x' \in X(H')$ such that $\forall (0 \subset K \subseteq N)$*:

$$\beta^*\langle x_K\rangle = \beta^*\langle x'_K\rangle.$$

Proof. Let H' be an rth-level descendant of H (i.e. $H' = P^{(r)}\ldots P^{(1)}H$). Using Theorem I r times, there exists x' such that for all K, $\beta^r\langle x'_K\rangle = \langle x_K\rangle$ Hence, $\beta^*\langle x'_K\rangle = \beta^*\langle x_K\rangle$.

THEOREM III. *If H' is a descendant of H, $\hat{R}_K(H') \supseteq \hat{R}_K(H)$. Hence, $\bigcap_{H \in \mathscr{C}(G)} \hat{R}_K(H) = R_K(G)$.*

Proof. From Lemma IV together with Theorem II we have $s \in \hat{R}_K(H) \Rightarrow s \in \hat{R}_K(H')$.

THEOREM IV. *$\bar{s} \in \Gamma_K(G) \Rightarrow \forall s_K \exists s_{N-K} : s \in M_K\bar{s}$.*

Proof. Let $\bar{s} \in \Gamma_K(G)$. By definition there exists $\bar{H} \in \mathscr{C}(G)$ such that $\bar{s} \in \hat{R}_K(\bar{H})$. Hence, from Lemma IV there exists $\bar{x}_{N-K} \in X_{N-K}(\bar{H})$ such that $\bar{s} \in \beta^*\langle \bar{x}_{N-K}\rangle \subseteq M_K\bar{s}$. Now suppose that, in contradiction to the theorem, $\exists s'_K \forall s_{N-K} : (s'_K, s_{N-K}) \in \tilde{M}_K\bar{s}$; that is, $\exists s'_K : \langle s'_K\rangle \subseteq \tilde{M}_K\bar{s}$. From Theorem II there exists $x'_K \in X_K(\bar{H})$ such that $\beta^*\langle x'_K\rangle = \langle s'_K\rangle$. Hence altogether there exist $\bar{x}_{N-K}$ and x'_K such that

$$\beta^*\langle \bar{x}_{N-K}\rangle \subseteq M_K\bar{s}; \quad \beta^*\langle x'_K\rangle \subseteq \tilde{M}_K\bar{s},$$

implying that $\beta^*\langle \bar{x}_{N-K}\rangle \cap \beta^*\langle x'_K\rangle = 0$. This, however, is impossible, since by Lemma II this intersection contains $\beta^*\langle \bar{x}_{N-K}, x'_K\rangle$, which is non-empty by Lemma I.

THEOREM V. $[\forall s_K \exists s_{N-K}: s \in M_K \bar{s}] \Rightarrow \bar{s} \in \Gamma_K(G)$. Hence, using Theorem IV, $\Gamma_K(G) = \{\bar{s} \mid \forall s_K \exists s_{N-K}: s \in M_K \bar{s}\}$.

Proof. Consider the prediction structure $P^{(N-K)}$ defined as in Section IV above. In the game $P^{(N-K)}G$, a strategy for a player $i \in K$ is effectively a choice of a basic strategy s_i, while a strategy for a player $j \in N-K$ is effectively a choice of a function $f_j \in S_j^{S_K}$. Hence, an $x \in X(P^{(N-K)}G)$ may be written $x = (s_i: i \in K; f_j: j \in N-K)$, with $\beta x = (s_i: i \in K; f_j(s_K): j \in N-K)$, or $x = (s_K, f_{N-K})$ with $\beta x = (s_K, f_{N-K}(s_K))$.

Now assume the antecedent of the theorem:

$$\forall s_K \exists s_{N-K}: s \in M_K \bar{s}.$$

This may (by the axiom of choice) be written

$$\exists f_{N-K} \forall s_K: (s_K, f_{N-K}(s_K)) \in M_K \bar{s}.$$

Take one of the f_{N-K} here asserted to exist and form the function $\vec{f}_{N-K}$ such that

$$\vec{f}_{N-K}(\bar{s}_K) = \bar{s}_{N-K}$$

and

$$\vec{f}_{N-K}(s_K) = f_{N-K}(s_K), \quad \text{for} \quad s_K \neq \bar{s}_K.$$

Then since $\bar{s} \in M_K \bar{s}$ we have, altogether

$$\exists \vec{f}_{N-K}: \begin{cases} \forall s_K: (s_K, \vec{f}_{N-K}(s_K)) \in M_K \bar{s}; \\ \vec{f}_{N-K}(\bar{s}_K) = \bar{s}_{N-K}. \end{cases}$$

But this can be written

$$\exists \bar{x}_{N-K}: \bar{s} \in \beta \langle \bar{x}_{N-K} \rangle \subseteq M_K \bar{s},$$

which proves the theorem.

THEOREM VI. $[\exists s_{N-K} \forall s_K: s \in M_K \bar{s}] \Rightarrow \bar{s} \in \Sigma_K(G)$.

Proof. The antecedent is

$$\exists s_{N-K}: \langle s_{N-K} \rangle \subseteq M_K \bar{s}$$

which from Theorem II implies

$$\forall H \exists x_{N-K}: \beta^* \langle x_{N-K} \rangle \subseteq M_K \bar{s}.$$

(Note. From now on we let H be a typical element of $\mathscr{C}(G)$ and let x be a typical outcome of the H last mentioned.)

This is the same as

$$\forall H \, \exists x_{N-K} \, \forall x_K : \beta^* x \in M_K \bar{s}.$$

This implies

$$\forall H \, \forall x_K \, \exists x_{N-K} : \beta^* x \in M_K \bar{s},$$

from which we can write, using Theorem II

$$\forall H \, \exists \bar{x} \begin{cases} \forall x_K \, \exists x_{N-K} : x \in M'_K \bar{x}; \\ \beta^* \bar{x} = \bar{s} \end{cases}$$

– where M'_K refers to preferences in the game H. From Theorem V, however, this is the same as

$$\forall H \, \exists \bar{x} : \begin{cases} \bar{x} \in \Gamma_K(H); \\ \beta^* \bar{x} = \bar{s}, \end{cases}$$

which proves the theorem, using the definition of $\Sigma_K(G)$ as $\bigcap_H \hat{\Gamma}_K(H)$.

THEOREM VII. $\bar{s} \in \Sigma_K(G) \Rightarrow \exists s_{K-N} \, \forall s_K : s \in M_K \bar{s}$.

Hence, using Theorem VI, $\Sigma_K(G) = \{\bar{s} \mid \exists s_{N-K} \, \forall s_K : s \in M_K \bar{s}\}$.

Proof. From Theorem V and the definition of $\Sigma_K(G)$ as $\bigcap_H \hat{\Gamma}_K(H)$, the antecedent implies:

$$\forall H : \forall x_K \, \exists x_{N-K} : \beta^* x \in M_K \bar{s}.$$

This assertion concerns all metagames $H \in \mathscr{C}(G)$. Consider in particular $P^{(K)}G$ as defined in Section IV. An outcome of this game effectively is an (f_K, s_{N-K}) with $\beta(f_K, s_{N-K}) = (f_K(s_{N-K}), s_{N-K})$. (See proof of Theorem V.) Hence, the antecedent of the theorem implies

$$\forall f_K \, \exists s_{N-K} : (f_K(s_{N-K}), s_{N-K}) \in M_K \bar{s}.$$

This is the same as

$$\sim \exists f_K \, \forall s_{N-K} : (f_K(s_{N-K}), s_{N-K}) \notin M_K \bar{s},$$

which implies

$$\sim \forall s_{N-K} \exists s_K : s \notin M_K \bar{s},$$

which is the same as the consequent of the theorem.

We have now proved all our theorems about metarational outcomes. The remaining theorems concern metaequilibria.

THEOREM VIII. *If H' is a descendant of H, $\hat{E}(H') \supseteq \hat{E}(H)$. Hence, $\bigcap_H \hat{E}(H) = E(G)$.*

Proof. From Lemma V together with Theorem II we have $\bar{s} \in \hat{E}(H) \Rightarrow$
$\Rightarrow \bar{s} \in \hat{E}(H')$.

THEOREM IX. $\Gamma(G) \subseteq \bigcap_{i \in N} \Gamma_i(G)$.

Proof. By definition, we have

$$\Gamma(G) = \bigcup_H \hat{E}(H) = \bigcup_H \beta^* \bigcap_i R_i(H).$$

This right-hand side is a subset of

$$\bigcup_H \bigcap_i \beta^* R_i(H) = \bigcup_H \bigcap_i \hat{R}_i(H) \subseteq \bigcap_i \bigcup_H \hat{R}_i(H) = \bigcap_i \Gamma_i(G).$$

THEOREM X. $[\forall i \, \forall s_i \, \exists s_{N-i} : s \in M_i \bar{s}] \Rightarrow \bar{s} \in \Gamma(G)$.

Hence, using Theorem IX and Theorem V, we have

$$\Gamma(G) = \bigcap_{i \in N} \Gamma_i(G) = \{\bar{s} \mid \forall_i \forall s_i \exists s_{N-i} : s \in M_i \bar{s}\}.$$

Proof. Assume the antecedent is true for some given $\bar{s}$. Then by the axiom of choice we have

$$\forall i \, \exists f_{N-i} \, \forall s_i : (s_i, f_{N-i}(s_i)) \in M_i \bar{s},$$

where

$$f_{N-i} : S_i \to S_{N-i}.$$

Now for each i, take one of the f_{N-i} here asserted to exist, write it f_{N-i}^i to denote its dependence on i, and finally write it as an indexed set $(f_j^i : j \in N-i)$ of functions $f_j^i : S_i \to S_j$.

Clearly, this indexed set obeys

$$\forall s_i : (s_i; \ f_j^i(s_i) : j \in N-i) \in M_i \bar{s}.$$

With the help of these indexed sets (one for each player i), I shall

'GENERAL' METAGAMES 279

construct a second-level metagame $P'PG$ such that $\bar{s} \in \hat{E}(P'PG)$, and so prove the theorem. To this end, let P be the prediction structure $(P_i : i \in N)$ such that, for each i, $P_i = \{\bar{p}_i, \hat{p}_i\}$, where $\bar{p}_i = \{\bar{s}_i\}$ and $\hat{p}_i = S_i$. Thus in PG each player i must either commit himself precisely to $\bar{s}_i$, or make only the empty commitment to S_i. A strategy (p_i, f_i) for player i in PG is thus either a choice of $p_i = \bar{p}_i$ (in which case f_i will be 'constant at' $\bar{s}_i$) or it is a choice of $p_i = \hat{p}_i$ together with a function

$$f_i : \prod_{j \in N-i} \{\bar{p}_j, \hat{p}_j\} \to S_i.$$

Next, let the structure P' be a prediction structure on PG such that $P'PG$ has the following extensive form. *Stage I.* Each player i follows a certain procedure in choosing a 'constraint' c_i constraining his choice of a (p_i, f_i) in the game PG. The procedure is as follows. Player i may choose a c_i, denoted $\bar{c}_i$, consisting of the statement: '$f_i(\bar{p}_{N-i}) = \bar{s}_i$'. Alternatively, player i may select any $s_i \in S_i$ and choose a c_i, denoted $c_i^{(s_i)}$, consisting of the statement

$$'p_i = \hat{p}_i, \text{ and } f_i(\hat{p}_{N-i}) = s_i'.$$

Stage II. Each player j, informed of the constraints chosen at Stage I, chooses a (p_j, f_j) within the constraint c_j that he himself chose.

It remains to prove that $\bar{s} \in \hat{E}(P'PG)$. To this end, consider the following strategy for player i in the game $P'PG$. *Stage I.* Choose $\bar{c}_i$. *Stage II.* (i) If the other players have chosen $\bar{c}_j$, $j \in N-i$, choose $p_i = \bar{p}_i$ (and consequently choose f_i 'constant at' $\bar{s}_j$). (ii) If for some unique $k \in N-i$ the remaining players $j \in N-i-k$ have chosen $\bar{c}_j$, while player k has chosen $c_k^{(s_k)}$, choose $p_i = \hat{p}_i$ and the function $f_i = f_i^{(s_k)}$ given by

$$f_i^{(s_k)}(\bar{p}_{N-i}) = \bar{s}_i;$$
$$f_i^{(s_k)}(p_{N-i}) = f_i^k(s_k), \text{ for } p_{N-i} \neq \bar{p}_{N-i},$$

where f_i^k is the function discussed at the start of this proof. (iii) In all other cases, choose (p_i, f_i) arbitrarily subject to '$f_i(\bar{p}_{N-i}) = \bar{s}_i$'.

Denote the above strategy in $P'PG$ by $\bar{\sigma}_i$, and write $\bar{\sigma} = (\bar{\sigma}_i : i \in N)$. Clearly, $\beta^2 \bar{\sigma} = \bar{s}$, since if in accordance with $\bar{\sigma}$ each i at Stage I chooses $\bar{c}_i$, $\bar{\sigma}$ requires each i at Stage II to choose $\bar{p}_i = \{\bar{s}_i\}$. Next, consider what any player k can achieve if each $i \in N-k$ chooses $\bar{\sigma}_i$, while k chooses otherwise. We consider two possibilities. (i) If k continues to choose $\bar{c}_k$, the others will make commitments $\{\bar{s}_i\}$, to which $\bar{c}_k$ dictates that k's response shall be $\bar{s}_k$. The result will be $\bar{s}$. (ii) Suppose that k chooses $c_k^{(s_k)}$ for some s_k. By assumption all the others have chosen constraints $\bar{c}_i$, so that he is revealed as the unique 'defaulter', and each other player i will choose the commitment $\hat{p}_i$ and the function $f_i^{(s_k)}$, resulting in the basic outcome

$$(f_k(\hat{p}_{N-k}); \; f_i^{(s_k)}(\hat{p}_{N-i}): i \in N-k).$$

The constraint $c_k^{(s_k)}$ determines the kth component of this, while the definitions of $(f_i^{(s_k)}: i \in N-k)$ determine the other components. Hence, the outcome is

$$(s_k; f_i^k(s_k): i \in N-k),$$

which was shown at the start of this proof to belong to $\bar{s}$.

We have shown that $\beta^2 \bar{\sigma} = \bar{s}$ and that $\beta^2 \langle \bar{\sigma}_{N-k} \rangle \subseteq M_k \bar{s}$. Hence, $\bar{s} \in \hat{E}(P'PG)$. This proves the theorem.

THEOREM XI. $\Sigma(G) = \bigcap_{i \in N} \Sigma_i(G)$.

Hence, using Theorem VII, $\Sigma(G) = \{\bar{s} \mid \forall i \, \exists s_{N-i} \, \forall s_i : s \in M_i \bar{s}\}$.

Proof. By definition,

$$\Sigma(G) = \bigcap_H \beta^* \bigcap_i \Gamma_i(H).$$

Now for any given metagame H, form the sets

$$\beta^* \bigcap_i \Gamma_i(H), \quad \bigcap_i \beta^* \Gamma_i(H).$$

One sees formally that the first is a subset of the second. I shall prove that the second is a subset of the first, so that the two are equal. Indeed, the condition for $\bar{s}$ to belong to the second is, from Theorem V,

$$\forall i \, \exists \bar{x} \begin{cases} \forall x_i \, \exists x_{N-i} : \beta^* x \in M_i \beta^* \bar{x}; \\ \beta^* \bar{x} = \bar{s}, \end{cases}$$

where $x, \bar{x} \in X(H)$. This implies

$$\forall i \, \forall x_i \, \exists x_{N-i} : \beta^* x \in M_i \bar{s},$$

which implies (using Theorem I)

$$\exists \bar{x} \begin{cases} \forall i\, \forall x_i\, \exists x_{N-i} : \beta^* x \in M_i \bar{s}; \\ \beta^* \bar{x} = \bar{s}, \end{cases}$$

or, equivalently,

$$\exists \bar{x} \begin{cases} \forall_i\, \forall x_i\, \exists x_{N-i} : \beta^* x \in M_i \beta^* \bar{x}; \\ \beta^* \bar{x} = \bar{s}, \end{cases}$$

which is the condition for $\bar{s}$ to belong to the first of my two expressions. Hence, we have

$$\Sigma(G) = \bigcap_H \bigcap_i \beta^* \Gamma_i(H) = \bigcap_i \bigcap_H \hat{\Gamma}_i(H) = \bigcap_i \Sigma_i(G),$$

by definition. Q.E.D.

I have now proved all that I set out to prove. I will add a theorem that assures the non-emptiness of $\Gamma(G)$ under rather general conditions. As defined in [2], a *partly ordinal* game G is a game in which each function M_i obeys not only the *reflexive* condition:

$$s \in M_i s, \quad \text{for all} \quad s;$$

but also the condition of *transitivity for the 'preferred' relation* $\tilde{M}_i$:

$$(s \in \tilde{M}_i s', \quad s' \in \tilde{M}_i s'') \Rightarrow (s \in \tilde{M}_i s'').$$

A *finite* game is one in which each set S_i is finite.

I now prove:

THEOREM XII. $\Gamma(G)$ is non-empty for any finite partly ordinal game G.

Proof. The theorem will be proved if we can show that for each player i there exists a strategy s_i^* such that

$$\forall s_{N-i} : (s_i^*, s_{N-i}) \in \Gamma_i,$$

for then, it will be the case that

$$s^* = (s_1^*, \ldots, s_n^*) \in \bigcap_i \Gamma_i = \Gamma.$$

But suppose that for some player i no such s_i^* exists. Then we have

$$\forall s_i \, \exists s_{N-i} : s \notin \Gamma_i$$

which is equivalent to

$$\exists \psi_{N-i} \, \forall s_i : (s_i, \psi_{N-i}(S_i)) \notin \Gamma_i$$

where

$$\psi_{N-i} : S_i \to S_{N-i}.$$

Select a function ψ_{N-i} such as is here required to exist. Now select an arbitrary $s_i = s_i^1$. Since $(s_i^1, \psi_{N-i}(s_i^1)) \notin \Gamma_i$, it is the case that

$$\exists s_i^2 \forall s_{N-i} : (s_i^2, s_{N-i}) \in \tilde{M}_i(s_i^1, \psi_{N-i}(s_i^1)),$$

which implies, in particular,

$$\exists s_i^2 : (s_i^2, \psi_{N-i}(s_i^2)) \in \tilde{M}_i(s_i^1, \psi_{N-i}(s_i^1)).$$

Next, $(s_i^2, \psi_{N-i}(s_i^2)) \notin \Gamma_i$. Hence,

$$\exists s_i^3 : (s_i^3, \psi_{N-i}(s_i^3)) \in \tilde{M}_i(s_i^2, \psi_{N-i}(s_i^2)).$$

Proceeding in this way we obtain an infinite sequence of strategies

$$s_i^1, s_i^2, s_i^3, \ldots,$$

such that

$$(s_i^2, \psi_{N-i}(s_i^2)) \in \tilde{M}_i(s_i^1, \psi_{N-i}(s_i^1))$$
$$(s_i^3, \psi_{N-i}(s_i^3)) \in \tilde{M}_i(s_i^2, \psi_{N-i}(s_i^2))$$
$$(s_i^4, \psi_{N-i}(s_i^4)) \in \tilde{M}_i(s_i^3, \psi_{N-i}(s_i^3)) \ldots$$

No two strategies in this infinite sequence can be the same. For if $t<r$, we have by the transitivity rule for $\tilde{M}_i$

$$(s_i^r, \psi_{N-i}(s_i^r)) \in \tilde{M}_i(s_i^t, \psi_{N-i}(s_i^t))$$

and by the reflexive condition we cannot have $s \in \tilde{M}_i s$. Hence, the set $\{s_i^1, s_i^2, s_i^3, \ldots\}$ is infinite, contradicting the assumption that S_i is finite. This proves the theorem.

To conclude, I state the following

META-THEOREM. *The preceding proofs remain valid if $\bar{\mathscr{C}}(G)$ is substituted for $\mathscr{C}(G)$.*

Dept. of Systems Design,
University of Waterloo, Ontario

BIBLIOGRAPHY

[1] Aumann, R. J., 'The Core of a Cooperative Game without Side Payments', *Transactions of the American Mathematical Society* **98** (1961) 539–552.
[2] Howard, N., *Paradoxes of Rationality: Theory of Metagames and Political Behaviour*, MIT Press, 1971.

THEORY AND DECISION LIBRARY

An International Series in the Philosophy and Methodology
of the Social and Behavioral Sciences

Editors:

GERALD EBERLEIN, *Universität des Saarlandes*
WERNER LEINFELLNER, *University of Nebraska*

1. GÜNTER MENGES (ed.), *Information, Inference, and Decision,* viii + 190 pp. (approx.)
2. ANATOL RAPOPORT (ed.), *Game Theory as a Theory of Conflict Resolution,* v + 283 pp.
3. MARIO BUNGE (ed.), *The Methodological Unity of Science,* viii + 264 pp.
4. COLIN CHERRY (ed.), *Pragmatic Aspects of Human Communication,* ix + 178 pp.